D1522871

Benchmark Papers on Energy

Series Editors:
R. Bruce Lindsay, Brown University
Mones E. Hawley, Professional Services International

PUBLISHED VOLUMES AND VOLUMES IN PREPARATION

ENERGY: Historical Development of the Concept / *R. Bruce Lindsay*
APPLICATIONS OF ENERGY: Nineteenth Century / *R. Bruce Lindsay*
COAL, PART I: Social, Economic, and Environmental Aspects / *Mones E. Hawley*
COAL, PART II: Scientific and Technical Aspects / *Mones E. Hawley*
THE SECOND LAW OF THERMODYNAMICS / *Joseph Kestin*
IRREVERSIBLE PROCESSES / *Joseph Kestin*
ATOMIC PHYSICS AND ENERGY / *Robert Lindsay*
ENERGY STORAGE / *W. V. Hassenzahl*
DYNAMICS AND CONTROL OF POWER SYSTEMS / *A. H. El-Abiad*

A BENCHMARK ® Books Series

THE SECOND LAW OF THERMODYNAMICS

Edited by
JOSEPH KESTIN
Brown University

Benchmark Papers on Energy, Volume 5
Library of Congress Catalog Card Number: 76-19059
ISBN: 0-87933-242-5

78 77 76 1 2 3 4 5
Manufactured in the United States of America.

LIBRARY OF CONGRESS CATALOGING IN PUBLICATION DATA
Main entry under title:
The Second law of thermodynamics.
(Benchmark papers in energy ; 5)
Includes index.
CONTENTS: The first steps: Carnot, S. Reflections on the motive power of fire and on machines fitted to develop that power. Clapeyron, E. Memoir on the motive power of heat. Thomson, W. (Lord Kelvin) On an absolute thermometric scale founded on Carnot's theory of the motive power of heat and calculated from Regnault's observations. [etc.]
1. Thermodynamics—Addresses, essays, lectures. I. Kestin, Joseph.
QC311.19.S4 536'.71 76-19059
ISBN 0-87933-242-5

Exclusive Distributor: **Halsted Press**
A Division of John Wiley & Sons, Inc.
ISBN: 0-470-98944-0

SERIES EDITOR'S FOREWORD

The Benchmark Papers on Energy constitute a series of volumes that makes available to the reader in carefully organized form important and seminal articles on the concept of energy, including its historical development, its applications in all fields of science and technology, and its role in civilization in general. This concept is generally admitted to be the most far-reaching idea that the human mind has developed to date, and its fundamental significance for human life and society is everywhere evident.

One group of volumes in the series contains papers bearing primarily on the evolution of the energy concept and its current applications in the various branches of science. Another group of volumes concentrates on the technological and industrial applications of the concept and its socio-economic implications.

Each volume has been organized and edited by an authority in the area to which it pertains and offers the editor's careful selection of the appropriate seminal papers, that is, those articles which have significantly influenced further development of that phase of the whole subject. In this way every aspect of the concept of energy is placed in proper perspective, and each volume represents an introduction and guide to further work.

Each volume includes an editorial introduction by the volume editor, summarizing the significance of the field being covered. Every article or group of articles is accompanied by editorial commentary, with explanatory notes where necessary. Both an author index and a subject index are provided for ready reference. Articles in languages other than English are either translated or summarized in English. It is the hope of the publisher and editor that these volumes will serve as a working library of the most important scientific, technological, and social literature connected with the idea of energy.

The Second Law of Thermodynamics is the fifth volume in the Energy Series. It has been prepared by Professor Joseph Kestin of the Division of Engineering at Brown University. In fifteen carefully selected papers the editor portrays the development of the second law and, to a large extent, the development of classical thermodynamics in general from the rather uncertain beginnings in the nineteenth century to the more rigorous formulation due to Planck and Carathéodory. An ample introduction and careful commentaries throw light on every aspect of the development. Stress is laid on the difficulties posed by irreversible processes, and this serves as a preview of Professor Kestin's following companion volume in the series, that on irreversible thermodynamics.

R. BRUCE LINDSAY

PREFACE

The second law of thermodynamics is one hundred and fifty-odd years old and yet it is still controversial. To be sure, there is not much real controversy when it is applied to the study of equilibrium states, even though some contemporary research workers favor a formalism in which the concept of thermodynamic equilibrium either has no place or plays an inconspicuous role. The problems begin when irreversible processes occupy the center of the stage.

This volume concentrates on the noncontroversial part of the second law and traces its development from Carnot via Clausius and Kelvin to Carathéodory. I think that I have chosen the most telling papers in this development, and that I did not miss anything of critical importance, but I am convinced that another editor would have made different choices. I apologize if I have offended the sensibility of some of my readers or of the specialist historians of this subject.

Even though I alone am to be blamed for potential omissions, I wish to express my appreciation to Professor R. Bruce Lindsay, the series editor, for his invitation to prepare this volume, for his continuing encouragement, and for the wise counsel that he was always ready to proffer. Thanks to him, I have spent many pleasant hours studying the old masters—the true discoverers of the second law.

Professor Lindsay went far beyond the call of duty as series editor with his superb translation of Clausius's "Abhandlung IX."

I also wish to express my thanks to Professor Josef Meixner, with whom I have been in the habit of discussing thermodynamics on and off for at least twenty years. He helped me to locate the works of Clausius and others, and through his writings—as evidenced by the last entry in this volume—he enabled me to form what appears to me to be a clear view of the subject.

My secretary, Leslie Comeau, patiently, uncomplainingly, and competently typed the several versions of the text and of both translations.

JOSEPH KESTIN

CONTENTS

CONTENTS BY AUTHOR

THE SECOND LAW OF THERMODYNAMICS

INTRODUCTION

Expressed in contemporary terms, the second law of thermodynamics consists of three statements. They all relate to equilibrium states of closed systems. (A closed system is defined as an arbitrary collection of matter enclosed by a boundary surface that is not itself crossed by matter. The system is not considered open if the boundary is crossed by electrons.)

The *first statement* (or first part of the second law) asserts that for any equilibrium state of a thermodynamic closed system it is possible to define a potential, that is, a function defined except for an additive constant, which is called entropy, S, and which is a function of state only. The most convenient canonical variables that define the state consist of the internal energy U and of a set α of deformation variables. Often the deformation variables are divided into a set of external deformation variables $\mathbf{a}$ (for example, the volume V of a fluid or the six independent components $\epsilon_1, \ldots, \epsilon_6$ of the symmetric strain tensor in an elastic solid) and a set of internal deformation variables ζ (for example, the extent ξ of a chemical reaction).

In other words, it is asserted that in the coordinate space $U, \mathbf{a}, \zeta$ there exists a single-valued potential $S(U, \mathbf{a}, \zeta)$ or, inverting, that there exists a fundamental equation

$$U = U(S, \mathbf{a}, \zeta) \qquad (1)$$

whose perfect differential is

$$dU = TdS - F_i da_i + A_j d\zeta_j \qquad \text{(sums)} \qquad (1a)$$

where

$$T = \left(\frac{\partial U}{\partial S}\right)_{\mathbf{a},\zeta} \qquad \text{is the temperature,} \tag{1b}$$

$$-F_i = \left(\frac{\partial U}{\partial a_i}\right)_{a_j,\zeta,S} \qquad \text{are the generalized forces,} \tag{1c}$$

and

$$A_j = \left(\frac{\partial U}{\partial \zeta_j}\right)_{\zeta_k,S,\mathbf{a}} \qquad \text{are the affinities.} \tag{1d}$$

Every point in the space $U, \mathbf{a}, \zeta$ or, equivalently, in $S, \mathbf{a}, \zeta$ represents an equilibrium state, and a line in it represents a reversible process (continuous sequence of equilibrium states). For any such reversible process, Eq. (1a) can be written in the form of a rate equation

$$\dot{U} = T\dot{S} - F_i\dot{a}_i + A_j\dot{\zeta}_j\,, \tag{2}$$

in which T, F_i, and A_j are specified functions of the canonical variables S, $\mathbf{a}$, ζ.

The preceding statement is equivalent to saying that the entropy difference $S_2 - S_1$ between two arbitrary equilibrium states of any system is calculated with the aid of the Clausius integral

$$S_2 - S_1 = \int_{\substack{1\\R}}^{2} \frac{dQ^\circ}{T}\,, \tag{3}$$

where dQ°, the heat exchanged in a reversible process (curve R in space U, $\mathbf{a}$, ζ), is

$$dQ^\circ = dU + F_j\,da_j - A_j\,d\zeta_j. \tag{4}$$

The curve R is arbitrary, and the thermodynamic temperature T constitutes one (the most convenient) of an infinity of integrating denominators of the linear differential (Pfaffian) form (4) that is evidently integrable.

The *second statement* (second part of the second law) asserts that during an arbitrary irreversible process (continuous sequence of nonequilibrium states) that starts with an equilibrium state 1 and ends in an equilibrium state 2 and that is also *adiabatic*, the entropy difference

$$S_2 - S_1 = \Theta, \tag{5}$$

known as the entropy production Θ, must be nonnegative, or

$$\Theta \geqslant 0, \tag{6}$$

the equality sign being valid only for reversible adiabatic processes.

Equivalently, the change in the entropy of a system that performs an irreversible process between equilibrium states 1 and 2 and exchanges quantities of heat Q_k with an arbitrary number of ideal heat reservoirs of temperatures T_k consists of two terms, of an entropy flux Q_k/T_k (sum) that can be of any sign (it is equal to zero during an adiabatic process) and a nonnegative entropy production term Θ, so that now

$$S_2 - S_1 = \frac{Q_k}{T_k} + \Theta. \tag{5a}$$

Thus the second statement can be expressed most generally by the inequality (6), that is, by the stipulation that the entropy production must be nonnegative in any process. The entropy production is defined as the difference in the sums of the entropies of the system and of all the heat sources (real or fictitious) interacting with it during the process.

The *third statement*, the equilibrium principle, asserts that the entropy S of an equilibrium state is a maximum with respect to all variations in state during which the values of U, a, ζ remain constant. The variations are thought of as reversible processes during which the system is subdivided into parts of different energies, U_α ($U = \Sigma U_\alpha$), different values of each individual deformation variable (e.g., different volumes V_β with $V = \Sigma V_\beta$), etc. Thus the maximum in S occurs in an enlarged space of states compared to U, a, ζ. The maximum entropy property assures the convexity of the hypersurface $S = S\ (U, \mathrm{a}, \zeta)$.

In this book we shall introduce the seminal papers that have led to the discovery of the preceding statements.

Note that the first part of the second law deals exclusively with equilibrium states of systems and establishes that a unique entropy can be assigned to each of them (disregarding the trivial fact that an arbitrary additive constant is contained in it). The fact that nonequilibrium states are not considered is not accidental. No unambiguously defined entropy of a nonequilibrium state has yet been discovered. In fact, we accept the view, first advanced by Meixner (Paper 15; see also Bataille and Kestin 1975) that a unique entropy cannot be assigned to a nonequilibrium state. In any case a systematic study of the thermodynamics of nonequilibrium states (Prigogine 1947, 1961; Meixner and Reik 1959; De Groot and Mazur 1962; Coleman and Noll 1963) has been undertaken only comparatively recently, has not yet been completed, and requires a separate volume to itself. It is planned to publish a Benchmark volume entitled *Irreversible Processes.*

The second part of the second law does not exclude irreversible processes, that is, sequences of nonequilibrium states, but insists that at least the initial and final state of the process must be states of equilibrium. This is because the mathematical statement is made in terms of entropy, a quantity that we indicate, calculate, and tabulate for equilibrium states only.

The discoveries that have led us to the acceptance of the preceding statements of the second law of thermodynamics have been made in small steps and over a span of nearly one hundred and fifty years, starting with the now-celebrated *Réflexions sur la puissance motrice du feu* published in 1824 by Sadi Carnot, "former student of the École Polytechnique."

When studying the original papers, it is necessary constantly to bear in mind that the authors were trying not only to develop physical statements regarding the efficiency of heat engines, the transformation of various energy forms into each other, and later, the relation between energy, heat, and work. They started with a clean slate and were compelled to develop, step by small step, the very concepts in terms of which clear statements could be made at all. As is usual, they meandered, and it is understandable that the essential concepts emerged quite late in this process of development, and that the contributors were not as keenly aware of the need to distinguish between ideal, equilibrium and real, nonequilibrium states of systems as we are now. Furthermore, they were often sidetracked into undertaking lines of investigation—such as the study of gas laws—that did not bear upon the second law directly but were essential for its reduction to statements that could be confronted with experiment.

Another problem that taxed the reasoning powers of the early contributors was the struggle clearly to formulate their understanding of, and the distinction between, such fundamental concepts as "heat," "work," and "energy" that we now employ and "force," "caloric," "fire," and others that have completely disappeared in their early, imprecise meanings in thermodynamics. This was connected with the philosophical considerations regarding "the nature of heat" and the role of "temperature" in thermodynamics. In particular, it must be remembered that a clear enunciation of the first law did not join the mainstream of science much before the middle of the nineteenth century, that is nearly 25 years after Sadi Carnot's "Réflexions." In this connection see R. Bruce Lindsay: *Applications of Energy: Nineteenth Century*, Benchmark Papers on Energy, Vol. 2.

To complicate matters further, the "caloric" theory of heat was so ingrained in the minds of the early workers that many rejected the principle of energy conservation. Thus the early progress in the development of the second law was achieved by scientists for whom the "caloric" theory was "obvious," and the problem of reconciling Carnot with Mayer and Joule loomed large in the early writings. For this reason many of their statements seem incorrect at first reading. To this one must also add the shortcomings of the mathematical notation used at the time, and the general lack of familiarity with the theory of linear differential forms.

In making my selections I concentrated on illustrating the development of *ideas* and *concepts* as distinct from *specific results.* Hence I put a good deal of emphasis on the earlier contributors—Carnot, Clapeyron, Lord Kelvin, and Clausius—and thus have placed a greater burden on the reader than would be the case otherwise. I was also, and quite understandably, unable to avoid a fair amount of repetitiveness because, as is usual, before a discovery is communicated to the reader, past errors must be corrected, and a new framework must be established.

The articles are divided into four parts corresponding to the four principal thrusts in the development of thermodynamics. Part I centers on the properties of reversible Carnot cycles studied in the context of the "caloric" theory of heat, except for Paper 5, in which Clausius, still hesitantly, introduces the first law in its modern form and enunciates the second law, albeit without due emphasis. Part II contains the essential parts of three papers, one by Lord Kelvin and two by Clausius, in which the second law and many of its consequences are presented in a form that still persists, essentially unchanged, in modern elementary textbooks. They contain the most mature presentation of the subject by its two codiscoverers and concentrate on equilibrium states even though they make no emphatic statements to this effect. The only irreversible process for which a mathematical expression is given is one that starts and ends in an equilibrium state. Part II concludes with two papers by Lord Kelvin in which he amplifies and applies his new discoveries. Part III contains Gibbs's extension to the study of stable equilibria and Carathéodory's mathematically elegant reformulation of the whole subject. The second half of this part contains two contributions, one by Max Planck and the other by Born, each of which illustrates several subsidiary points.

Max Planck's contribution is the chapter on the second law from his *Theory of Heat*, which was translated into a number of languages, gave excellent service through many editions, and was regarded as a bible by generations of theoretical and applied physicists. The chapter clearly illustrates the concept of inaccessible states and the role of thermodynamic temperature as an integrating denominator for the Pfaffian of heat in a reversible process. Both these ideas were utilized by Carathéodory.

The last contribution, Max Born's Lecture V (with appendixes), in his famous, sweeping review of "cause" and "chance"—the Waynflete lectures—closely follows his paper that appeared in *Physikalische Zeitschrift* in 1921, and contains a compact and lucid *précis* of thermostatics in the idiom of Carathéodory, created at Born's instigation in the first place.

Part IV consists of a single paper by Meixner; in it the author clearly recognizes the limitations imposed by the requirement that states 1 and 2 in Eqs. (5) and (5a) *must* be equilibrium states and indicates the need to erect a true science of nonequilibrium states on the foundations laid so far. This provides us with a gateway into the volume *Irreversible Processes.*

REFERENCES

Bataille, J., and J. Kestin. 1975. L'interprétation physique de la thermodynamique rationelle. *J. Mécanique,* 14:365–384.

Coleman, B. D., and W. Noll. 1963. The Thermodynamics of Elastic Materials with Heat Conduction and Viscosity. *Arch. Rational Mech. Anal.,* 13:167–178.

De Groot, S. R., and P. Mazur. 1962. *Non-Equilibrium Thermodynamics.* North-Holland, Amsterdam.

Meixner, J., and H. G. Reik. 1959. Thermodynamik der irreversiblen Prozesse. *Encyclopedia of Physics,* Vol. III/2, S. Fluegge. Springer-Verlag, Berlin. pp. 413–523.

Prigogine, I. 1947. *Étude thermodynamique des phénomènes irréversibles.* Desoer, Liège.

——. 1961. *Thermodynamics of Irreversible Processes.* Interscience Publishers, New York, London.

SELECTED BIBLIOGRAPHY OF THERMOSTATICS

Buchdahl, H. A. 1966. *Concepts of Classical Thermodynamics.* Cambridge University Press, Cambridge.

Callen, H. B. 1960. *Thermodynamics.* Wiley & Sons, New York.

Epstein, P. S. 1937. *Textbook of Thermodynamics.* Wiley & Sons, New York.

Guggenheim, E. A. 1957. *Thermodynamics.* North-Holland, Amsterdam.

Kestin, J. 1966. *A Course in Thermodynamics.* Blaisdell, Waltham, Toronto, London.

Kestin, J. 1968. *A Course in Thermodynamics,* Vol. 2. Blaisdell, Waltham, Toronto, London.

Kestin, J., and J. R. Dorfman. 1971. Chapter 1: Summary of Classical Thermodynamics. In *A Course in Statistical Thermodynamics*, pp. 3–26. Academic Press, New York, London.

Kirchhoff, G. 1894. *Vorlesungen über Theorie der Wärme.* Teubner, Leipzig.

Landau, L. D., and E. M. Lifshitz. 1969. *Statistical Physics.* 2nd ed. J. B. Sykes, and M. J. Kearsley, trans., Pergamon Press, Oxford, and Addison-Wesley, Reading, Mass.

Pippard, A. B. 1957. *The Elements of Classical Thermodynamics.* Cambridge University Press, Cambridge.

Planck, M. 1932. *Introduction to Theoretical Physics, Vol. V: Theory of Heat.* H. L. Brose, trans., Macmillan and Co., London. (Originally published as *Vorlesungen über Thermodynamik*, Veit and Co., Leipzig, 1897.)

Poincaré, H. 1892. *Thermodynamique.* Georges Carré, Paris.

Prigogine, I., and R. Defay. 1954. *Chemical Thermodynamics.* D. H. Everett, trans., Longmans Green and Co., London.

Reiss, H. 1965. *Methods of Thermodynamics.* Blaisdell, Waltham, Toronto, London.

Schottky, W. 1929. *Thermodynamik.* Springer, Berlin.

Sommerfeld, A. 1956. *Lectures on Theoretical Physics, Vol. V: Thermodynamics and Statistical Mechanics*, J. Kestin, trans., Academic Press, New York.

Tisza, L. 1966. *Generalized Thermodynamics.* The M.I.T. Press, Cambridge, Mass.

Wilson, A. H. 1957. *Thermodynamics and Statistical Mechanics.* Cambridge University Press, Cambridge.

Zemansky, M. W. 1957. *Heat and Thermodynamics.* 5th ed. McGraw-Hill, New York.

Part I

THE FIRST STEPS

Editor's Comments on Papers 1 Through 5

It is now well known that the second law of thermodynamics "started with Sadi Carnot"; yet a reader who is not versed in the subject would be hard put to see the connection between the modern statements given in our earlier Introduction and the extract contained in Paper 1. Indeed a long and creative chain of discoveries separates the two.

Sadi Carnot set out to formulate the physical principles that could be employed to understand and improve the operation of the first practical prime mover known to history—the reciprocating steam engine. As we now know (Klein 1974), in fulfilling this task he attempted to do for heat engines what his father, the French revolutionary general Lazare Carnot, accomplished for strictly mechanical machines, namely, to formulate by powerful reasoning a number of abstract principles for the maximization of their efficiency. It is these principles, transformed as they have become by his successors, that form the true precursors of the second law.

It is not possible to understand Carnot's work on the second law without spending some time on his ideas concerning the first law (Lindsay 1975, 1976). We now accept the fact, first clearly recognized by Kelvin and Clausius, that the statements of the two laws are logically independent. Nevertheless, when reading Carnot and his early followers, one's clarity of comprehension is impeded by the fact that they adopted the material, or "caloric," theory of heat. Using the water turbine/water pump as an analog example, Carnot supposed that the quantity of heat absorbed was exactly equal to the quantity of heat rejected by the steam engine. Many historians of science could not reconcile themselves to the fact that neither Sadi Carnot nor Sir William Thomson (Lord Kelvin) and Rudolf Clausius in their early writings understood the first law as we understand it today, and engaged in puny attempts to whitewash them from the vantage point of hindsight. For example, on p. 7 (**20**), Carnot states clearly: "The steam is here only a means of transporting the caloric" and "the production of motive power is then due . . . not to an actual consumption of caloric, but *to its transportation from a warm body to a cold body* . . . ." Similarly, Kelvin states on p. 102 (**54**) of Paper 3: ". . . the conversion of heat (or *caloric*) into mechanical effect is probably impossible, certainly undiscovered," adding in a footnote that "a contrary opinion however has been advocated by Mr. Joule of Manchester."

This collection of five papers presents to the reader four whose authors firmly believed the caloric theory of heat at the time of writing, and one, that due to Clausius, who was the first one to note that Carnot's discoveries were not necessarily contradictory to the principle of the equivalence of heat and work. Commenting on Kelvin, he says on p. 3 (**89**):

> I believe, nevertheless, that we ought not to suffer ourselves to be daunted by these difficulties; but that, on the contrary, we must look steadfastly into this theory which calls heat a motion, as in this way alone we can arrive at the means of establishing it or refuting it. Besides this, I do not imagine that the difficulties are so great as Thomson considers them to be.

In short, Clausius still had doubts or, in Kelvin's words [footnote on p. 102 (**54**)]: ". . . it must be confessed that as yet much is involved in mystery with reference to these fundamental questions of natural philosophy."

Instead of trying to impute to the great pioneer thermodynamicists thoughts they did not think, we may pause to admire their genius in being able to take such giant steps with respect to the second law in spite of their early adherence to the caloric theory of heat.

Carnot's great achievements started with his recognition that *two* heat reservoirs were the minimum necessary to produce power from heat, and that given two reservoirs, *maximum* power is secured when operating in a *cycle* that, moreover, must be *reversible.* This discovery led him to the natural abstract design of what we now call a Carnot engine. He then proceeded to show that the operation of such a reversible Carnot engine was independent of the working fluid employed in it, and that finally, the efficiency, or ratio of "motive power" to the "caloric," depended exclusively on the *temperatures* of the two heat reservoirs. He deduced these properties from the (to him) intuitively obvious principle that it should not be possible to construct, even in one's mind, a perpetual motion engine, no matter how ingenious the design. This may be taken as the first verbal formulation of the second law.

The reasoning that led him to this conclusion has survived, with modifications, to this day in many textbooks; it consists in coupling two reversible Carnot engines, one of which is made to work in the opposite sense from the other. Since the heat transfers of the pair cancel, it makes no difference to the conclusions whether the heat absorbed by one engine is or is not equal to that rejected by it.

Carnot's work remained unnoticed and unknown for a long time. So much so that Kelvin, who worked in Paris in Regnault's laboratory a quarter of a century after the publication of the *Réflexions*, was unable to obtain a copy for some time. In fact, he first became aware of Carnot's discoveries from the work of Clapeyron (Paper 2).

Émile Clapeyron recognized the importance of Carnot's work as well as the connection between it and the then vigorous research into the properties of gases and vapors. He invented the now so popular thermodynamic diagrams, depicted the Carnot cycle for a gas and for a saturated vapor in what we would call a p,v diagram, and recognized that the area enclosed by the curve of the cycle represented work. He further proceeded to apply mathematical analysis in which, as a graduate of the École de Mines, he was well educated. He invented the idea of an elementary cycle, cycle *abcd* in Figure 1 (which uses the later concept of entropy and thermodynamic temperature). By a reasoning that we would now describe as unduly complex, he was able to translate Carnot's argument to the equation

$$\frac{dW}{Q} = \frac{dt}{C(t)}. \tag{1}$$

He realized that Carnot's function $C(t)$ of the empirical temperature t

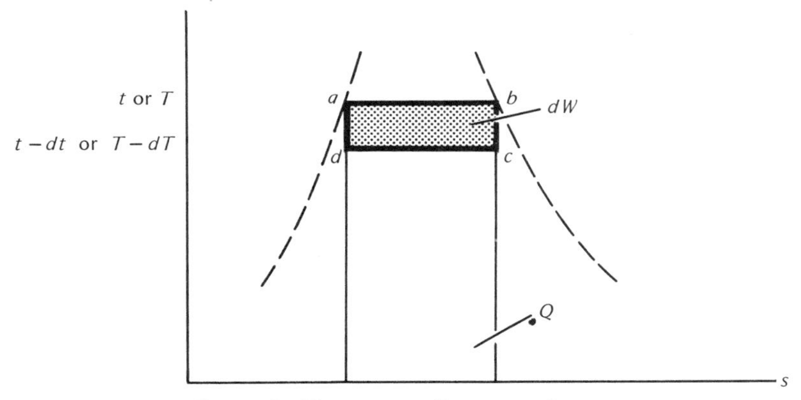

Figure 1 Elementary Carnot cycle

must be determined with reference to experiments, and he proceeded to do so numerically using the then known properties of gases and condensing vapors. Kelvin and Clausius performed similar calculations later.

When Eq. (1) is used to calculate $C(t)$ from the properties of condensed gases, in effect use is made of what we now call the Clausius–Clapeyron equation. This is because the latter is a consequence of Eq. (1) if points a and b of Figure 1 are placed on the vapor-pressure curve (shown sketeched in broken lines).

It is reasonable to suppose that infinitesimal Carnot cycles were used at this stage as a basis, because the early thermodynamicists had insufficient knowledge of the properties of substances to permit them to calculate over finite temperature intervals. Incidentally, in this limit the heat rejected tends to become equal to the heat absorbed, and the argument becomes identical in the context of a caloric theory of heat with that conducted on the basis of energy conservation.

The idea that the reversible Carnot cycle—that is, that effectively, Eq. (1) can be used to *define* the absolute thermodynamic temperature scale—came surprisingly early to Lord Kelvin (Paper 3), who, as already mentioned, knew Carnot's work at that stage only secondhand through his reading of Clapeyron's paper.

Whereas Clapeyron's paper was largely expository, Kelvin's contained an important discovery, namely, that

$$T = C(t) \tag{2}$$

can be used to define a new, absolute scale that is "independent of the properties of any particular kind of matter." He defined this scale by postulating "that a unit of heat descending from a body A at temperature $T°$ of this scale to a body B at the temperature $(T - 1)°$

would give out the same mechanical effect, whatever the number T." Let us denote this T by θ and observe that then

$$\frac{dW}{Q} = d\theta \tag{3}$$

defines θ. Alternatively, as Kelvin proposed in a later paper, we may define another absolute scale (and we do nowadays) by putting

$$\frac{dW}{Q} = \frac{dT}{T} \tag{4}$$

so that

$$d\theta \equiv d\ln T. \tag{5}$$

Thus Kelvin's first scale θ is a linear function of the logarithm of our absolute scale, the detailed relation still depending on the number and nature of the fixed thermometric points employed.

A contemporary author would start with Eq. (4) as a definition and would relegate the problem of calculating the relation

$$C(t) = T \tag{6}$$

to thermometry. More precisely, the historical C is given by $C = T/J$, where J is the mechanical equivalent of heat. Kelvin preferred to operate with the inverse $\mu(t) = J/T$. We interpret this as the problem of expressing the theoretically justified scale T in terms of an arbitrary, empirical scale t.

> The labor of performing the necessary calculations for effecting a comparison of the proposed scale with that of the air-thermometer, between the limits 0° and 230° of the latter has been kindly undertaken by Mr. William Steele, lately of Glasgow College, now of St. Peter's College, Cambridge. His results in tabulated forms were laid before the Society, with a diagram, in which the comparison between the two scales is represented graphically [Kelvin, Paper 3; part not reprinted here].

Lord Kelvin realized that the most satisfactory empirical temperature scale t was the one defined by a gas thermometer.

Since the operation of a reversible Carnot engine ("perfect thermo-dynamic engine" in Lord Kelvin's words) is independent of the working fluid used in it, all early contributors engaged in numerical calculations of $C(t)$ that required them to undertake a thorough study of the properties of gases and vapors, particularly if they wanted to—as they eventually had to—reason with Carnot cycles of finite extent. At this point the study of the properties of systems became inextricably entwined with progress in the understanding of the second law. Even though Clapeyron already had written down the equation of state of a perfect gas [in the form $pv = R(267 + t)$], and even though Clausius (in Paper 5) perceived that this was an asymptotic form and almost introduced the concept of a perfect gas [on p. 7 (**93**)], the fact that T becomes identical with the empirical temperature t of a perfect-gas thermometer remained unnoticed. The inclusion of the factor J in $C(t) = T/J$ undoubtedly made it more difficult to perceive the relation.

The last two papers in this section, Paper 4 by Kelvin and Paper 5 by Clausius, neatly illustrate the transition from the study of the second law in the context of the "caloric" theory of heat to that in "motional" theory. Kelvin accepts the "caloric" theory; Clausius clearly enunciates the first law in a form acceptable to us to this day.

Reading both of them, and bearing Carnot's original writings in mind, we become convinced that the new discoveries in the field of heat were considered "difficult" and in need of clarification and reorganization. Examined from our own perspective they still appear difficult, and they require study rather than reading to achieve understanding because we can identify the "blocks" to understanding contained in them that, though not bearing directly on the second law, impeded further progress. Not the least of them was rooted in the contemporary notation for derivatives and differentials.

To avoid repetition, we have excised most of the mathematical derivations from Clausius's paper, even though readers with a historical curiosity may find them interesting (Clausius 1851). We retained Kelvin's because they adequately illustrate the point and demonstrate how far it was possible to go in terms of the "caloric" theory of heat.

REFERENCES

Clausius, R. 1851. On the Moving Force of Heat, and the Laws Regarding the Nature of Heat Itself Which are Deducible Therefrom. *Phil. Mag.* Ser. 4, **2**:1–21, 102–119.

Klein, M. J. 1974. Carnot's Contribution to Thermodynamics. *Physics Today*, **27**(8):23–28.

Lindsay, R. B. 1975. *Energy: Historical Development of the Concept*. Benchmark Papers on Energy, Vol. 1. Dowden, Hutchinson & Ross, Stroudsburg, Pa.

Lindsay, R. B. 1976. *Applications of Energy: Nineteenth Century*. Benchmark Papers on Energy, Vol. 2. Dowden, Hutchinson & Ross, Stroudsburg, Pa.

1

Reprinted by permission of the publisher from pp. 3–22 of
Reflections on the Motive Power of Fire and Other Papers, E. Mendoza, ed.,
Dover Publications, Inc., New York, 1960, 174 pp.

Reflections on the Motive Power of Fire, and on Machines Fitted to Develop that Power

S. Carnot

EVERY one knows that heat can produce motion. That it possesses vast motive-power no one can doubt, in these days when the steam-engine is everywhere so well known.

To heat also are due the vast movements which take place on the earth. It causes the agitations of the atmosphere, the ascension of clouds, the fall of rain and of meteors, the currents of water which channel the surface of the globe, and of which man has thus far employed but a small portion. Even earthquakes and volcanic eruptions are the result of heat.

From this immense reservoir we may draw the moving force necessary for our purposes. Nature, in providing us with combustibles on all sides, has given us the power to produce, at all times and in all places, heat and the impelling power which is the result of it. To develop this power, to appropriate it to our uses, is the object of heat-engines.

The study of these engines is of the greatest interest, their importance is enormous, their use is continually increasing, and they seem destined to produce a great revolution in the civilized world.

Already the steam-engine works our mines, impels our ships, excavates our ports and our rivers, forges iron, fashions wood, grinds grains, spins and weaves our cloths, transports the heaviest burdens, etc. It appears that it must some day serve as a universal motor, and be substituted for animal power, waterfalls, and air currents.

Over the first of these motors it has the advantage of economy, over the two others the inestimable advantage that it can be used at all times and places without interruption.

If, some day, the steam-engine shall be so perfected that it can be set up and supplied with fuel at small cost, it will combine all desirable qualities, and will afford to the industrial arts a range the extent of which can scarcely be predicted. It is not merely that a powerful and convenient motor that can be procured and carried anywhere is substituted for the motors already in use, but that it causes rapid extension in the arts in which it is applied, and can even create entirely new arts.

The most signal service that the steam-engine has rendered to England is undoubtedly the revival of the working of the coal mines, which had declined, and threatened to cease entirely, in consequence of the continually increasing difficulty of drainage, and of raising the coal.* We should rank second the benefit to iron manufacture, both by the abundant supply of coal substituted for wood just when the latter had begun to grow scarce, and by the powerful machines of all kinds, the use of which the introduction of the steam-engine has permitted or facilitated.

Iron and heat are, as we know, the supporters, the bases, of the mechanic arts. It is doubtful if there be in England a single industrial establishment of which the existence does not depend on the use of these agents, and which does not freely employ them. To take away today from England her steam-engines would be to take away at the same time her coal and iron. It would be to dry up all her sources of wealth, to ruin all on which her prosperity depends, in short, to annihilate that colossal power. The destruction of her navy, which she considers her strongest defence, would perhaps be less fatal.

The safe and rapid navigation by steamships may be regarded as an entirely new art due to the steam-engine. Already this art has permitted the establishment of prompt and regular communications across the arms of the sea, and on the great rivers of the old and new continents. It has made it possible to traverse savage regions where before we could scarcely penetrate. It has enabled us to carry the fruits of civilization over portions of the globe where they would else have been wanting for years. Steam navigation brings nearer together the most distant nations. It tends to unite the nations of the earth as inhabitants of one country. In fact, to lessen the time,

* It may be said that coal-mining has increased tenfold in England since the invention of the steam-engine. It is almost equally true in regard to the mining of copper, tin, and iron. The results produced in a half-century by the steam-engine in the mines of England are today paralleled in the gold and silver mines of the New World—mines of which the working declined from day to day, principally on account of the insufficiency of the motors employed in the draining and the extraction of the minerals.

the fatigues, the uncertainties, and the dangers of travel—is not this the same as greatly to shorten distances?*

The discovery of the steam-engine owed its birth, like most human inventions, to rude attempts which have been attributed to different persons, while the real author is not certainly known. It is, however, less in the first attempts that the principal discovery consists, than in the successive improvements which have brought steam-engines to the conditions in which we find them today. There is almost as great a distance between the first apparatus in which the expansive force of steam was displayed and the existing machine, as between the first raft that man ever made and the modern vessel.

If the honor of a discovery belongs to the nation in which it has acquired its growth and all its developments, this honor cannot be here refused to England. Savery, Newcomen, Smeaton, the famous Watt, Woolf, Trevithick, and some other English engineers, are the veritable creators of the steam-engine. It has acquired at their hands all its successive degrees of improvement. Finally, it is natural that an invention should have its birth and especially be developed, be perfected, in that place where its want is most strongly felt.

Notwithstanding the work of all kinds done by steam-engines, notwithstanding the satisfactory condition to which they have been brought today, their theory is very little understood, and the attempts to improve them are still directed almost by chance.

The question has often been raised whether the motive power of heat† is unbounded, whether the possible improvements in steam-engines have an assignable limit—a limit which the nature of things will not allow to be passed by any means whatever; or whether, on the contrary, these improvements may be carried on indefinitely. We have long sought, and are seeking today, to ascertain whether there are in existence agents preferable to the vapor of water for developing the motive power of heat; whether atmospheric air, for

---

* We say, to lessen the dangers of journeys. In fact, although the use of the steam-engine on ships is attended by some danger which has been greatly exaggerated, this is more than compensated by the power of following always an appointed and well-known route, of resisting the force of the winds which would drive the ship towards the shore, the shoals, or the rocks.

† We use here the expression motive power to express the useful effect that a motor is capable of producing. This effect can always be likened to the elevation of a weight to a certain height. It has, as we know, as a measure, the product of the weight multiplied by the height to which it is raised.

example, would not present in this respect great advantages. We propose now to submit these questions to a deliberate examination.

The phenomenon of the production of motion by heat has not been considered from a sufficiently general point of view. We have considered it only in machines the nature and mode of action of which have not allowed us to take in the whole extent of application of which it is susceptible. In such machines the phenomenon is, in a way, incomplete. It becomes difficult to recognize its principles and study its laws.

In order to consider in the most general way the principle of the production of motion by heat, it must be considered independently of any mechanism or any particular agent. It is necessary to establish principles applicable not only to steam-engines* but to all imaginable heat-engines, whatever the working substance and whatever the method by which it is operated.

Machines which do not receive their motion from heat, those which have for a motor the force of men or of animals, a waterfall, an air current, etc., can be studied even to their smallest details by the mechanical theory. All cases are foreseen, all imaginable movements are referred to these general principles, firmly established, and applicable under all circumstances. This is the character of a complete theory. A similar theory is evidently needed for heat-engines. We shall have it only when the laws of physics shall be extended enough, generalized enough, to make known beforehand all the effects of heat acting in a determined manner on any body.

We will suppose in what follows at least a superficial knowledge of the different parts which compose an ordinary steam-engine; and we consider it unnecessary to explain what are the furnace, boiler, steam-cylinder, piston, condenser, etc.

The production of motion in steam-engines is always accompanied by a circumstance on which we should fix our attention. This circumstance is the re-establishing of equilibrium in the caloric; that is, its passage from a body in which the temperature is more or less elevated, to another in which it is lower. What happens in fact in a steam-engine actually in motion? The caloric developed in the furnace by the effect of the combustion traverses the walls of the boiler, produces steam, and in some way incorporates itself with it.

* We distinguish here the steam-engine from the heat-engine in general. The latter may make use of any agent whatever, of the vapor of water or of any other, to develop the motive power of heat.

The latter carrying it away, takes it first into the cylinder, where it performs some function, and from thence into the condenser, where it is liquefied by contact with the cold water which it encounters there. Then, as a final result, the cold water of the condenser takes possession of the caloric developed by the combustion. It is heated by the intervention of the steam as if it had been placed directly over the furnace. The steam is here only a means of transporting the caloric. It fills the same office as in the heating of baths by steam, except that in this case its motion is rendered useful.

We easily recognize in the operations that we have just described the re-establishment of equilibrium in the caloric, its passage from a more or less heated body to a cooler one. The first of these bodies, in this case, is the heated air of the furnace; the second is the condensing water. The re-establishment of equilibrium of the caloric takes place between them, if not completely, at least partially, for on the one hand the heated air, after having performed its function, having passed round the boiler, goes out through the chimney with a temperature much below that which it had acquired as the effect of combustion; and on the other hand, the water of the condenser, after having liquefied the steam, leaves the machine with a temperature higher than that with which it entered.

The production of motive power is then due in steam-engines not to an actual consumption of caloric, but *to its transportation from a warm body to a cold body*, that is, to its re-establishment of equilibrium—an equilibrium considered as destroyed by any cause whatever, by chemical action such as combustion, or by any other. We shall see shortly that this principle is applicable to any machine set in motion by heat.

According to this principle, the production of heat alone is not sufficient to give birth to the impelling power: it is necessary that there should also be cold; without it, the heat would be useless. And in fact, if we should find about us only bodies as hot as our furnaces, how can we condense steam? What should we do with it if once produced? We should not presume that we might discharge it into the atmosphere, as is done in some engines;* the atmosphere would not receive it. It does receive it under the actual condition of things, only because it fulfils the office of a vast

* Certain engines at high pressure throw the steam out into the atmosphere instead of the condenser. They are used specially in places where it would be difficult to procure a stream of cold water sufficient to produce condensation.

condenser, because it is at a lower temperature; otherwise it would soon become fully charged, or rather would be already saturated.*

Wherever there exists a difference of temperature, wherever it has been possible for the equilibrium of the caloric to be re-established, it is possible to have also the production of impelling power. Steam is a means of realizing this power, but it is not the only one. All substances in nature can be employed for this purpose, all are susceptible of changes of volume, of successive contradictions and dilatations, through the alternation of heat and cold. All are capable of overcoming in their changes of volume certain resistances, and of thus developing the impelling power. A solid body—a metallic bar for example—alternately heated and cooled increases and diminishes in length, and can move bodies fastened to its ends. A liquid alternately heated and cooled increases and diminishes in volume, and can overcome obstacles of greater or less size, opposed to its dilatation. An aeriform fluid is susceptible of considerable change of volume by variations of temperature. If it is enclosed in an expansible space, such as a cylinder provided with a piston, it will produce movements of great extent. Vapors of all substances capable of passing into a gaseous condition, as of alcohol, of mercury, of sulphur, etc., may fulfil the same office as vapor of water. The latter, alternately heated and cooled, would produce motive power in the shape of permanent gases, that is, without ever returning to a liquid state. Most of these substances have been proposed, many even have been tried, although up to this time perhaps without remarkable success.

We have shown that in steam-engines the motive-power is due to a re-establishment of equilibrium in the caloric; this takes place not only for steam-engines, but also for every heat-engine—that is,

---

* The existence of water in the liquid state here necessarily assumed, since without it the steam-engine could not be fed, supposes the existence of a pressure capable of preventing this water from vaporizing, consequently of a pressure equal or superior to the tension of vapor at that temperature. If such a pressure were not exerted by the atmospheric air, there would be instantly produced a quantity of steam sufficient to give rise to that tension, and it would be necessary always to overcome this pressure in order to throw out the steam from the engines into the new atmosphere. Now this is evidently equivalent to overcoming the tension which the steam retains after its condensation, as effected by ordinary means.

If a very high temperature existed at the surface of our globe, as it seems certain that it exists in its interior, all the waters of the ocean would be in a state of vapor in the atmosphere, and no portion of it would be found in a liquid state.

for every machine of which caloric is the motor. Heat can evidently be a cause of motion only by virtue of the changes of volume or of form which it produces in bodies.

These changes are not caused by uniform temperature, but rather by alternations of heat and cold. Now to heat any substance whatever requires a body warmer than the one to be heated; to cool it requires a cooler body. We supply caloric to the first of these bodies that we may transmit it to the second by means of the intermediary substance. This is to re-establish, or at least to endeavor to re-establish, the equilibrium of the caloric.

It is natural to ask here this curious and important question: Is the motive power of heat invariable in quantity, or does it vary with the agent employed to realize it as the intermediary substance, selected as the subject of action of the heat?

It is clear that this question can be asked only in regard to a given quantity of caloric,* the difference of the temperatures also being given. We take, for example, one body A kept at a temperature of 100° and another body B kept at a temperature of 0°, and ask what quantity of motive power can be produced by the passage of a given portion of caloric (for example, as much as is necessary to melt a kilogram of ice) from the first of these bodies to the second. We inquire whether this quantity of motive power is necessarily limited, whether it varies with the substance employed to realize it, whether the vapor of water offers in this respect more or less advantage than the vapor of alcohol, of mercury, a permanent gas, or any other substance. We will try to answer these questions, availing ourselves of ideas already established.

We have already remarked upon this self-evident fact, or fact which at least appears evident as soon as we reflect on the changes of volume occasioned by heat: *wherever there exists a difference of temperature, motive power can be produced.* Reciprocally, wherever we can consume this power, it is possible to produce a difference of temperature, it is possible to occasion destruction of equilibrium in the caloric. Are not percussion and the friction of bodies actually means of raising their temperature, of making it reach spontaneously a

* It is considered unnecessary to explain here what is quantity of caloric or quantity of heat (for we employ these two expressions indifferently), or to describe how we measure these quantities by the calorimeter. Nor will we explain what is meant by latent heat, degree of temperature, specific heat, etc. The reader should be familiarized with these terms through the study of the elementary treatises of physics or of chemistry.

higher degree than that of the surrounding bodies, and consequently of producing a destruction of equilibrium in the caloric, where equilibrium previously existed? It is a fact proved by experience, that the temperature of gaseous fluids is raised by compression and lowered by rarefaction. This is a sure method of changing the temperature of bodies, and destroying the equilibrium of the caloric as many times as may be desired with the same substance. The vapor of water employed in an inverse manner to that in which it is used in steam-engines can also be regarded as a means of destroying the equilibrium of the caloric. To be convinced of this we need to observe closely the manner in which motive power is developed by the action of heat on vapor of water. Imagine two bodies *A* and *B*, kept each at a constant temperature, that of *A* being higher than that of *B*. These two bodies, to which we can give or from which we can remove the heat without causing their temperatures to vary, exercise the functions of two unlimited reservoirs of caloric. We will call the first the furnace and the second the refrigerator.

If we wish to produce motive power by carrying a certain quantity of heat from the body *A* to the body *B* we shall proceed as follows:*

(1) To borrow caloric from the body *A* to make steam with it—that is, to make this body fulfil the function of a furnace, or rather of the metal composing the boiler in ordinary engines—we here assume that the steam is produced at the same temperature as the body *A*.

(2) The steam having been received in a space capable of expansion, such as a cylinder furnished with a piston, to increase the volume of this space, and consequently also that of the steam. Thus rarefied, the temperature will fall spontaneously, as occurs with all elastic fluids; admit that the rarefaction may be continued to the point where the temperature becomes precisely that of the body *B*.

(3) To condense the steam by putting it in contact with the body *B*, and at the same time exerting on it a constant pressure until it is entirely liquefied. The body *B* fills here the place of the injection-water in ordinary engines, with this difference, that it condenses the vapor without mingling with it, and without changing its own temperature.†

---

* [This is only a sketch and Carnot accidentally leaves the cycle incomplete. E. M.]

† We may perhaps wonder here that the body *B* being at the same temperature as the steam is able to condense it. Doubtless this is not strictly possible, but the slightest difference of temperature will determine the condensation, which suffices

The operations which we have just described might have been performed in an inverse direction and order. There is nothing to prevent forming vapor with the caloric of the body *B*, and at the temperature of that body, compressing it in such a way as to make it acquire the temperature of the body *A*, finally condensing it by contact with this latter body, and continuing the compression to complete liquefaction.

By our first operations there would have been at the same time production of motive power and transfer of caloric from the body *A* to the body *B*. By the inverse operations there is at the same time expenditure of motive power and return of caloric from the body *B* to the body *A*. But if we have acted in each case on the same quantity of vapor, if there is produced no loss either of motive power or caloric, the quantity of motive power produced in the first place will be equal to that which would have been expended in the second, and the quantity of caloric passed in the first case from the body *A* to the body *B* would be equal to the quantity which passes back again in the second from the body *B* to the body *A*; so that an indefinite number of alternative operations of this sort could be carried on without in the end having either produced motive power or transferred caloric from one body to the other.

Now if there existed any means of using heat preferable to those which we have employed, that is, if it were possible by any method whatever to make the caloric produce a quantity of motive power greater than we have made it produce by our first series of operations, it would suffice to divert a portion of this power in order by the method just indicated to make the caloric of the body *B* return

---

to establish the justice of our reasoning. It is thus that, in the differential calculus, it is sufficient that we can conceive the neglected quantities indefinitely reducible in proportion to the quantities retained in the equations, to make certain of the exact result.

The body *B* condenses the steam without changing its own temperature—this results from our supposition. We have admitted that this body may be maintained at a constant temperature. We take away the caloric as the steam furnishes it. This is the condition in which the metal of the condenser is found when the liquefaction of the steam is accomplished by applying cold water externally, as was formerly done in several engines. Similarly, the water of a reservoir can be maintained at a constant level if the liquid flows out at one side as it flows in at the other.

One could even conceive the bodies *A* and *B* maintaining the same temperature, although they might lose or gain certain quantities of heat. If, for example, the body *A* were a mass of steam ready to become liquid, and the body *B* a mass of ice ready to melt, these bodies might, as we know, furnish or receive caloric without thermometric change.

to the body *A* from the refrigerator to the furnace, to restore the initial conditions, and thus to be ready to commence again an operation precisely similar to the former, and so on: this would be not only perpetual motion, but an unlimited creation of motive power without consumption either of caloric or of any other agent whatever. Such a creation is entirely contrary to ideas now accepted, to the laws of mechanics and of sound physics. It is inadmissible.* We should then conclude that *the maximum of motive power resulting from the employment of steam is also the maximum of motive power realizable by any means whatever.* We will soon give a second more vigorous demonstration of this theory. This should be considered only as an approximation. (See page 15.)

We have a right to ask, in regard to the proposition just enunciated, the following questions: What is the sense of the word *maximum* here? By what sign can it be known that this maximum is attained? By what sign can it be known whether the steam is employed to greatest possible advantage in the production of motive power?

Since every re-establishment of equilibrium in the caloric may be the cause of the production of motive power, every re-establish-

---

* The objection may perhaps be raised here, that perpetual motion, demonstrated to be impossible by mechanical action alone, may possibly not be so if the power either of heat or electricity be exerted; but is it possible to conceive the phenomena of heat and electricity as due to anything else than some kind of motion of the body, and as such should they not be subjected to the general laws of mechanics? Do we not know besides, *à posteriori*, that all the attempts made to produce perpetual motion by any means whatever have been fruitless?—that we have never succeeded in producing a motion veritably perpetual, that is, a motion which will continue forever without alteration in the bodies set to work to accomplish it? The electromotor apparatus (the pile of Volta) has sometimes been regarded as capable of producing perpetual motion; attempts have been made to realize this idea by constructing dry piles said to be unchangeable; but however it has been done, the apparatus has always exhibited sensible deteriorations when its action has been sustained for a time with any energy.

The general and philosophic acceptation of the words *perpetual motion* should include not only a motion susceptible of indefinitely continuing itself after a first impulse received, but the action of an apparatus, of any construction whatever, capable of creating motive power in unlimited quantity, capable of starting from rest all the bodies of nature if they should be found in that condition, of overcoming their inertia; capable, finally, of finding in itself the forces necessary to move the whole universe, to prolong, to accelerate incessantly, its motion. Such would be a veritable creation of motive power. If this were a possibility, it would be useless to seek in currents of air and water or in combustibles this motive power. We should have at our disposal an inexhaustible source upon which we could draw ta will.

ment of equilibrium which shall be accomplished without production of this power should be considered as an actual loss. Now, very little reflection would show that all change of temperature which is not due to a change of volume of the bodies can be only a useless re-establishment of equilibrium in the caloric.* The necessary condition of the maximum is, then, *that in the bodies employed to realize the motive power of heat there should not occur any change of temperature which may not be due to a change of volume.* Reciprocally, every time that this condition is fulfilled the maximum will be attained. This principle should never be lost sight of in the construction of heat-engines; it is its fundamental basis. If it cannot be strictly observed, it should at least be departed from as little as possible.

Every change of temperature which is not due to a change of volume or to chemical action (an action that we provisionally suppose not to occur here) is necessarily due to the direct passage of the caloric from a more or less heated body to a colder body. This passage occurs mainly by the contact of bodies of different temperatures; hence such contact should be avoided as much as possible. It cannot probably be avoided entirely, but it should at least be so managed that the bodies brought in contact with each other differ as little as possible in temperature. When we just now supposed, in our demonstration, the caloric of the body *A* employed to form steam, this steam was considered as generated at the temperature of the body *A*; thus the contact took place only between bodies of equal temperatures; the change of temperature occurring afterwards in the steam was due to dilatation, consequently to a change of volume. Finally, condensation took place also without contact of bodies of different temperatures. It occurred while exerting a constant pressure on the steam brought in contact with the body *B* of the same temperature as itself. The conditions for a maximum are thus found to be fulfilled. In reality the operation cannot proceed exactly as we have assumed. To determine the passage of caloric from one body to another, it is necessary that there should be an excess of temperature in the first, but this excess may be supposed as slight as we please. We can regard it as insensible in theory, without thereby destroying the exactness of the arguments.

* We assume here no chemical action between the bodies employed to realize the motive power of heat. The chemical action which takes place in the furnace is, in some sort, a preliminary action—an operation destined not to produce immediately motive power, but to destroy the equilibrium of the caloric, to produce a difference of temperature which may finally give rise to motion.

A more substantial objection may be made to our demonstration, thus: When we borrow caloric from the body *A* to produce steam, and when this steam is afterwards condensed by its contact with the body *B*, the water used to form it, and which we considered at first as being of the temperature of the body *A*, is found at the close of the operation at the temperature of the body *B*. It has become cool. If we wish to begin again an operation similar to the first, if we wish to develop a new quantity of motive power with the same instrument, with the same steam, it is necessary first to re-establish the original condition—to restore the water to the original temperature. This can undoubtedly be done by at once putting it again in contact with the body *A*; but there is then contact between bodies of different temperatures, and loss of motive power.* It would be impossible to execute the inverse operation, that is, to return to the body *A* the caloric employed to raise the temperature of the liquid.

This difficulty may be removed by supposing the difference of temperature between the body *A* and the body *B* indefinitely small. The quantity of heat necessary to raise the liquid to its former temperature will be also indefinitely small and unimportant relatively to that which is necessary to produce steam—a quantity always limited.

The proposition found elsewhere demonstrated for the case in which the difference between the temperatures of the two bodies is indefinitely small, may be easily extended to the general case. In fact, if it operated to produce motive power by the passage of caloric from the body *A* to the body *Z*, the temperature of this latter body being very different from that of the former, we should imagine a series of bodies *B*, *C*, *D* . . . of temperatures intermediate between those of the bodies *A*, *Z*, and selected so that the differences from *A* to *B*, from *B* to *C*, etc., may all be indefinitely small. The caloric coming from *A* would not arrive at *Z* till after it had passed through

* This kind of loss is found in all steam-engines. In fact, the water destined to feed the boiler is always cooler than the water which it already contains. There occurs between them a useless re-establishment of equilibrium of caloric. We are easily convinced, *à posteriori*, that this re-establishment of equilibrium causes a loss of motive power if we reflect that it would have been possible to previously heat the feed-water by using it as condensing-water in a small accessory engine, when the steam drawn from the large boiler might have been used, and where the condensation might be produced at a temperature intermediate between that of the boiler and that of the principal condenser. The power produced by the small engine would have cost no loss of heat, since all that which had been used would have returned into the boiler with the water of condensation.

the bodies B, C, D, etc., and after having developed in each of these stages maximum motive power. The inverse operations would here be entirely possible, and the reasoning of page 11 would be strictly applicable.

According to established principles at the present time, we can compare with sufficient accuracy the motive power of heat to that of a waterfall. Each has a maximum that we cannot exceed, whatever may be, on the one hand, the machine which is acted upon by the water, and whatever, on the other hand, the substance acted upon by the heat. The motive power of a waterfall depends on its height and on the quantity of the liquid; the motive power of heat depends also on the quantity of caloric used, and on what may be termed, on what in fact we will call, the *height of its fall*,* that is to say, the difference of temperature of the bodies between which the exchange of caloric is made. In the waterfall the motive power is exactly proportional to the difference of level between the higher and lower reservoirs. In the fall of caloric the motive power undoubtedly increases with the difference of temperature between the warm and the cold bodies; but we do not know whether it is proportional to this difference. We do not know, for example, whether the fall of caloric from 100 to 50 degrees furnishes more or less motive power than the fall of this same caloric from 50 to zero. It is a question which we propose to examine hereafter.

We shall give here a second demonstration of the fundamental proposition enunciated on page 12, and present this proposition under a more general form than the one already given.

When a gaseous fluid is rapidly compressed its temperature rises. It falls, on the contrary, when it is rapidly dilated. This is one of the facts best demonstrated by experiment. We will take it for the basis of our demonstration.†

---

* The matter here dealt with being entirely new, we are obliged to employ expressions not in use as yet, and which perhaps are less clear than is desirable.

† The experimental facts which best prove the change of temperature of gases by compression or dilatation are the following:

(1) The fall of the thermometer placed under the receiver of a pneumatic machine in which a vacuum has been produced. This fall is very sensible on the Bréguet thermometer: it may exceed 40° or 50°. The mist which forms in this case seems to be due to the condensation of the watery vapor caused by the cooling of the air.

(2) The igniting of German tinder in the so-called pneumatic tinderboxes; which are, as we know, little pump-chambers in which the air is rapidly compressed.

If, when the temperature of a gas has been raised by compression, we wish to reduce it to its former temperature without subjecting its volume to new changes, some of its caloric must be removed. This caloric might have been removed in proportion as pressure was applied, so that the temperature of the gas would remain constant. Similarly, if the gas is rarefied we can avoid lowering the temperature by supplying it with a certain quantity of caloric. Let us call the caloric employed at such times, when no change of temperature occurs, *caloric due to change of volume*. This denomination does not indicate that the caloric appertains to the volume: it does not appertain to it any more than to pressure, and might as well be called *caloric due to the change of pressure*. We do not know what laws it follows relative to the variations of volume: it is possible that its quantity changes either with the nature of the gas, its density,

---

(3) The fall of a thermometer placed in a space where the air has been first compressed and then allowed to escape by the opening of a cock.

(4) The results of experiments on the velocity of sound. M. de Laplace has shown that, in order to secure results accurately by theory and computation, it is necessary to assume the heating of the air by sudden compression.

The only fact which may be adduced in opposition to the above is an experiment of MM. Gay-Lussac and Welter, described in the *Annales de Chimie et de Physique*. A small opening having been made in a large reservoir of compressed air, and the ball of a thermometer having been introduced into the current of air which passes out through this opening, no sensible fall of the temperature denoted by the thermometer has been observed.

Two explanations of this fact may be given: (1) The striking of the air against the walls of the opening by which it escapes may develop heat in observable quantity. (2) The air which has just touched the ball of the thermometer possibly takes again by its collision with this ball, or rather by the effect of the *détour* which it is forced to make by its rencounter, a density equal to that which it had in the receiver—much as the water of a current rises against a fixed obstacle, above its level.

The change of temperature occasioned in the gas by the change of volume may be regarded as one of the most important facts of physics, because of the numerous consequences which it entails, and at the same time as one of the most difficult to illustrate, and to measure by decisive experiments. It seems to present in some respects singular anomalies.

Is it not to the cooling of the air by dilatation that the cold of the higher regions of the atmosphere must be attributed? The reasons given heretofore as an explanation of this cold are entirely insufficient; it has been said that the air of the elevated regions receiving little reflected heat from the earth, and radiating towards celestial space, would lose caloric, and that this is the cause of its cooling; but this explanation is refuted by the fact that, at an equal height, cold reigns with equal and even more intensity on the elevated plains than on the summit of the mountains, or in those portions of the atmosphere distant from the sun.

or its temperature. Experiment has taught us nothing on this subject. It has only shown us that this caloric is developed in greater or less quantity by the compression of the elastic fluids.

This preliminary idea being established, let us imagine an elastic fluid, atmospheric air for example, shut up in a cylindrical vessel, *abcd* (Fig. 1), provided with a movable diaphragm or piston, *cd*. Let there be also two bodies, *A* and *B*, kept each at a constant temperature, that of *A* being higher than that of *B*. Let us picture to ourselves now the series of operations which are to be described:*

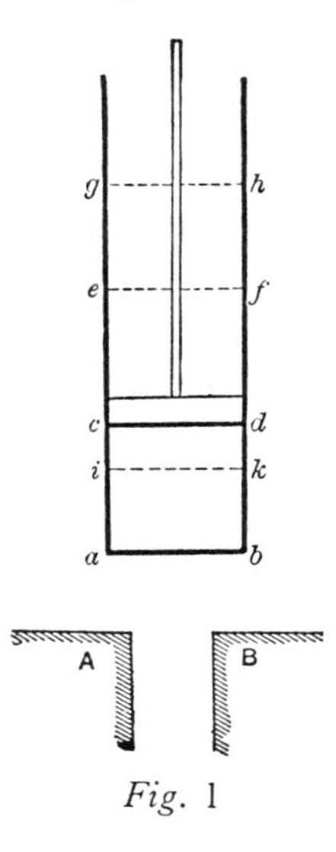

Fig. 1

(1) Contact of the body *A* with the air enclosed in the space *abcd* or with the wall of this space—a wall that we will suppose to transmit the caloric readily. The air becomes by such contact of the same temperature as the body *A*; *cd* is the actual position of the piston.

(2) The piston gradually rises and takes the position *ef*. The body *A* is all the time in contact with the air, which is thus kept at a constant temperature during the rarefaction. The body *A* furnishes the caloric necessary to keep the temperature constant.

(3) The body *A* is removed, and the air is then no longer in contact with any body capable of furnishing it with caloric. The

* ["Caloric" may be taken to mean "entropy." E. M.]

piston meanwhile continues to move, and passes from the position *ef* to the position *gh*. The air is rarefied without receiving caloric, and its temperature falls. Let us imagine that it falls thus till it becomes equal to that of the body *B*; at this instant the piston stops, remaining at the position *gh*.

(4) The air is placed in contact with the body *B*; it is compressed by the return of the piston as it is moved from the position *gh* to the position *cd*. This air remains, however, at a constant temperature because of its contact with the body *B*, to which it yields its caloric.

(5) The body *B* is removed, and the compression of the air is continued, which being then isolated, its temperature rises. The compression is continued till the air acquires the temperature of the body *A*. The piston passes during this time from the position *cd* to the position *ik*.

(6) The air is again placed in contact with the body *A*. The piston returns from the position *ik* to the position *ef*; the temperature remains unchanged.

(7) The step described under number (3) is renewed, then successively the steps (4), (5), (6), (3), (4), (5), (6), (3), (4), (5); and so on.

In these various operations the piston is subject to an effort of greater or less magnitude, exerted by the air enclosed in the cylinder; the elastic force of this air varies as much by reason of the changes in volume as of changes of temperature. But it should be remarked that with equal volumes, that is, for the similar positions of the piston, the temperature is higher during the movements of dilatation than during the movements of compression. During the former the elastic force of the air is found to be greater, and consequently the quantity of motive power produced by the movements of dilatation is more considerable than that consumed to produce the movements of compression. Thus we should obtain an excess of motive power—an excess which we could employ for any purpose whatever. The air, then, has served as a heat-engine; we have, in fact, employed it in the most advantageous manner possible, for no useless re-establishment of equilibrium has been effected in the caloric.

All the above-described operations may be executed in an inverse sense and order. Let us imagine that, after the sixth period, that is to say the piston having arrived at the position *ef*, we cause it to return to the position *ik*, and that at the same time we keep the air in contact with the body *A*. The caloric furnished by this body during the sixth period would return to its source, that is, to the body

A, and the conditions would then become precisely the same as they were at the end of the fifth period. If now we take away the body *A*, and if we cause the piston to move from *ik* to *cd*, the temperature of the air will diminish as many degrees as it increased during the fifth period, and will become that of the body *B*. We may evidently continue a series of operations the inverse of those already described. It is only necessary under the same circumstances to execute for each period a movement of dilatation instead of a movement of compression, and reciprocally.

The result of these first operations has been the production of a certain quantity of motive power and the removal of caloric from the body *A* to the body *B*. The result of the inverse operations is the consumption of the motive power produced and the return of the caloric from the body *B* to the body *A*; so that these two series of operations annul each other, after a fashion, one neutralizing the other.

The impossibility of making the caloric produce a greater quantity of motive power than that which we obtained from it by our first series of operations, is now easily proved. It is demonstrated by reasoning very similar to that employed at page 11; the reasoning will here be even more exact. The air which we have used to develop the motive power is restored at the end of each cycle of operations exactly to the state in which it was at first found, while, as we have already remarked, this would not be precisely the case with the vapor of water.*

We have chosen atmospheric air as the instrument which should develop the motive power of heat, but it is evident that the reasoning would have been the same for all other gaseous substances, and even for all other bodies susceptible of change of temperature through

* We tacitly assume in our demonstration, that when a body has experienced any changes, and when after a certain number of transformations it returns to precisely its original state, that is, to that state considered in respect to density, to temperature, to mode of aggregation—let us suppose, I say, that this body is found to contain the same quantitiy of heat that it contained at first, or else that the quantities of heat absorbed or set free in these different transformations are exactly compensated. This fact has never been called in question. It was first admitted without reflection, and verified afterwards in many cases by experiments with the calorimeter. To deny it would be to overthrow the whole theory of heat to which it serves as a basis. For the rest, we may say in passing, the main principles on which the theory of heat rests require the most careful examination. Many experimental facts appear almost inexplicable in the present state of this theory.

successive contractions and dilatations, which comprehends all natural substances, or at least all those which are adapted to realize the motive power of heat. Thus we are led to establish this general proposition:

The motive power of heat is independent of the agents employed to realize it; its quantity is fixed solely by the temperatures of the bodies between which is effected, finally, the transfer of the caloric.

We must understand here that each of the methods of developing motive power attains the perfection of which it is susceptible. This condition is found to be fulfilled if, as we remarked above, there is produced in the body no other change of temperature than that due to change of volume, or, what is the same thing in other words, if there is no contact between bodies of sensibly different temperatures.

Different methods of realizing motive power may be taken, as in the employment of different substances, or in the use of the same substance in two different states—for example, of a gas at two different densities.

This leads us naturally to those interesting researches on the aeriform fluids—researches which lead us also to new results in regard to the motive power of heat, and give us the means of verifying, in some particular cases, the fundamental proposition above stated.*

We readily see that our demonstration would have been simplified by supposing the temperatures of the bodies A and B to differ very little. Then the movements of the piston being slight during the periods (3) and (5), these periods might have been suppressed without influencing sensibly the production of motive power. A very little change of volume should suffice in fact to produce a very slight change of temperature, and this slight change of volume may be neglected in presence of that of the periods (4) and (6), of which the extent is unlimited.

If we suppress periods (3) and (5), in the series of operations above described, it is reduced to the following:

(1) Contact of the gas confined in *abcd* (Fig. 2) with the body A, passage of the piston from *cd* to *ef*.

(2) Removal of the body A, contact of the gas confined in *abef* with the body B, return of the piston from *ef* to *cd*.

---

* We will suppose, in what follows, the reader to be *au courant* with the later progress of modern physics in regard to gaseous substances and heat.

(3) Removal of the body *B*, contact of the gas with the body *A*, passage of the piston from *cd* to *ef*, that is, repetition of the first period, and so on.

The motive power resulting from the *ensemble* of operations (1) and (2) will evidently be the difference between that which is produced by the expansion of the gas while it is at the temperature of the body *A*, and that which is consumed to compress this gas while it is at the temperature of the body *B*.

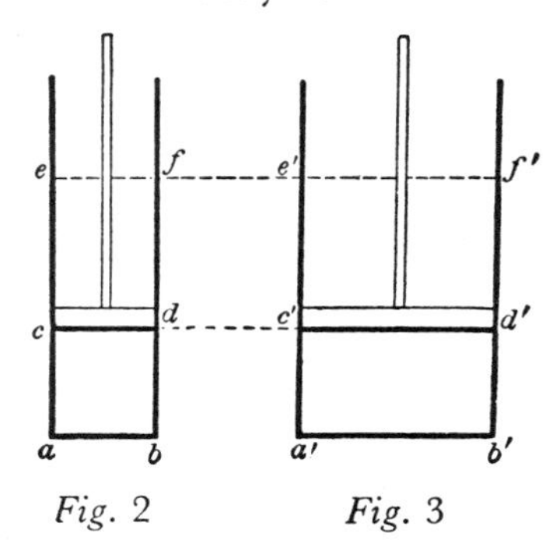

Fig. 2 *Fig.* 3

Let us suppose that operations (1) and (2) be performed on two gases of different chemical natures but under the same pressure—under atmospheric pressure, for example. These two gases will behave exactly alike under the same circumstances, that is, their expansive forces, originally equal, will remain always equal, whatever may be the variations of volume and of temperature, provided these variations are the same in both. This results obviously from the laws of Mariotte and MM. Gay-Lussac and Dalton—laws common to all elastic fluids, and in virtue of which the same relations exist for all these fluids between the volume, the expansive force, and the temperature.

Since two different gases at the same temperature and under the same pressure should behave alike under the same circumstances, if we subjected them both to the operations above described, they should give rise to equal quantities of motive power.

Now this implies, according to the fundamental proposition that we have established, the employment of two equal quantities of caloric; that is, it implies that the quantity of caloric transferred from the body *A* to the body *B* is the same, whichever gas is used.

The quantity of caloric transferred from the body *A* to the body *B* is evidently that which is absorbed by the gas in its expansion of volume, or that which this gas relinquishes during compression. We are led, then, to establish the following proposition:

When a gas passes without change of temperature from one definite volume and pressure to another volume and another pressure equally definite, the quantity of caloric absorbed or relinquished is always the same, whatever may be the nature of the gas chosen as the subject of the experiment.

Take, for example, 1 liter of air at the temperature of 100° and under the pressure of one atmosphere. If we double the volume of this air and wish to maintain it at the temperature of 100°, a certain quantity of heat must be supplied to it. Now this quantity will be precisely the same if, instead of operating on the air, we operate upon carbonic-acid gas, upon nitrogen, upon hydrogen, upon vapor of water or of alcohol, that is, if we double the volume of 1 liter of these gases taken at the temperature of 100° and under atmospheric pressure.

It will be the same thing in the inverse sense if, instead of doubling the volume of gas, we reduce it one half by compression. The quantity of heat that the elastic fluids set free or absorb in their changes of volume has never been measured by any direct experiment, and doubtless such an experiment would be very difficult, but there exists a datum which is very nearly its equivalent. This has been furnished by the theory of sound. It deserves much confidence because of the exactness of the conditions which have led to its establishment.

[*Editor's Note:* Material has been omitted at this point.]

2

Reprinted by permission of the publisher from pp. 73–88 of
Reflections on the Motive Power of Fire and Other Papers, E. Mendoza, ed.,
Dover Publications, Inc., New York, 1960, 174 pp.

Memoir on the Motive Power of Heat

E. Clapeyron

§ I

THERE are few questions more worthy of the attention of theoreticians and physicists than those which refer to the constitution of gases and vapors; the part which they play in nature, and the uses that industry puts them to, explain the many important investigations on them; but this vast question is far from being exhausted. Mariotte's law and Gay-Lussac's law state the relations which exist between the volume, pressure and temperature of a given quantity of gas; both have been accepted by scientists for a long time. The recent experiments carried out by Arago and Dulong leave no doubt as to the exactness of the first between very wide limits of pressure; but these results give no indication of the quantity of heat possessed by the gases or set free from them by pressure or lowering of temperature, neither do they give the law of specific heats at constant pressure and constant volume. This part of the theory of heat has nevertheless been the subject of careful investigations, among which may be mentioned the work of Delaroche and Bérard on the specific heats of gases. Lastly, in a memoir which he published under the title *Researches on the Specific Heats of Elastic Fluids*, Dulong established, by experiments which are free from all objections, that *equal volumes of all elastic fluids at a given temperature and pressure, compressed or expanded suddenly by a given fraction of their volumes, release or absorb the same absolute quantity of heat.**

Laplace and Poisson have published remarkable theoretical studies on this subject which however are based on hypotheses which can be disputed; they agree that the ratio of the specific heat at constant volume to that at constant pressure does not vary, and that the quantities of heat absorbed by gases are proportional to their temperatures.

Among studies which have appeared on the theory of heat I will mention finally a work by S. Carnot, published in 1824, with the

* [On the caloric theory, the heat set free when a given gas was compressed depended only on the initial and final volumes; it was the same whether the compression was isothermal or adiabatic. E. M.]

title *Reflections on the Motive Power of Fire*. The idea which serves as a basis of his researches seems to me to be both fertile and beyond question; his demonstrations are founded on *the absurdity of the possibility of creating motive power or heat out of nothing.* Here are statements of several theorems to which this new method of reasoning leads.

1. *When a gas passes, without changing its temperature, from a given volume and pressure to another given volume and pressure, the quantity of caloric absorbed or set free is always the same whatever the nature of the gas chosen for the experiment.*

2. *The difference between the specific heats under constant pressure and constant volume is the same for all gases.*

3. *When a gas varies its volume, without changing its temperature, the quantities of heat absorbed or set free by the gas are in arithmetic progression if the increases or decreases of volume are in geometric progression.*

This new method of demonstration seems to me worthy of the attention of theoreticians; it seems to me to be free of all objection, and it has acquired a new importance since the verification found in the work of Dulong, who has demonstrated experimentally the first theorem whose enunciation I have just recalled.

I believe that it is of some interest to take up this theory again; S. Carnot, avoiding the use of mathematical analysis, arrives by a chain of difficult and elusive arguments at results which can be deduced easily from a more general law which I shall attempt to prove. But before starting on the subject it is useful to restate the fundamental axiom which serves as a basis for Carnot's researches, and which will also be my starting point.

§II

It has been known for a long time that heat can develop motive power and, conversely, that by means of motive power heat can be produced. In the first case, it must be observed that a certain quantity of caloric always passes from one body at a given temperature to another body at a lower temperature; thus in steam engines, the production of mechanical force is accompanied by the passage of a part of the heat of combustion developed in the furnace whose temperature is very high, to the water of the condenser, whose temperature is very much lower.

Conversely it is always possible to use the passage of caloric from a hot body to a cold one to produce a mechanical force; to do this it is only necessary to construct a mechanism like an ordinary steam-engine, where the hot body generates the steam and the cold body acts as condenser.

The result is that there is a loss of *vis viva*, of mechanical force or quantity of action, whenever there is an immediate contact between two bodies at different temperatures and heat passes straight from one to another; therefore, in any mechanism designed to produce motive power from heat, there is a loss of force whenever there is a direct communication of heat between two bodies at different temperatures, and it follows that the maximum effect can be produced only by a mechanism in which contact is made only between bodies at equal temperatures.

Now our knowledge of the theory of gases and vapors shows how this can be achieved.

Let us imagine two bodies, one maintained at a temperature T, the other at a lower temperature t, such as for example the walls of a boiler in which the heat developed by combustion continually replaces that which the steam takes away; and the condenser of an ordinary heat-engine, in which a current of cold water all the time removes the heat given up by the steam by condensation and that due to its own temperature. For simplicity we will call the first body A, the second B.

Then, let us take any gas whatever at temperature T and let us put it in contact with the source A of heat; let us represent its volume v_0 by the abscissa AB and its pressure by the ordinate CB (Fig. 1).

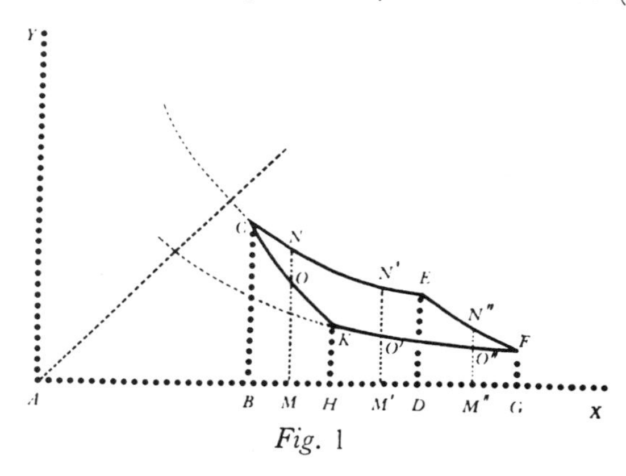

Fig. 1

If the gas is enclosed in a deformable vessel and is allowed to expand in an empty space where it can lose heat neither by radiation nor by contact, the source A of heat will at all times provide the quantity of caloric which its increase of volume causes to become latent, and it will keep the same temperature T. Its pressure however will diminish following Mariotte's law. The law of this variation can be represented geometrically by the curve CE where the abscissae are the volumes, and the ordinates the corresponding pressures.

Let us suppose that the expansion of the gas is continued till the original volume AB has become AD; and let DE be the pressure corresponding to the new volume; the gas will have developed a quantity of mechanical action during its expansion given by the integral of the product of the pressure times the differential of the volume, represented geometrically by the area contained between the axis of abscissae, the two co-ordinates CB, DE, and the portion CE of the hyperbola.

Let us suppose now that the body A is removed and that the expansion of the gas continues inside an envelope impermeable to heat; then since a part of its perceptible caloric becomes latent its temperature drops and its pressure decreases more rapidly according to an unknown law, which can be represented geometrically by a curve EF whose abscissae are the volume of the gas and whose ordinates are the corresponding pressures; we will suppose that the expansion of the gas is continued till the successive reductions of the perceptible caloric of the gas have brought it from the temperature T of the body A to the temperature t of the body B. Its volume is therefore AG, and the corresponding pressure FG.

It will be seen that the gas, during this second part of its expansion, develops a quantity of mechanical action represented by the area of the mixtilinear trapezium $DEFG$.

Now that the gas has been brought to the temperature t of the body B, let us bring the two into contact; if the gas is compressed in an envelope impermeable to heat, but in contact with the body B, the temperature of the gas tends to rise because of the release of latent caloric made perceptible by the compression, but as it is produced it is absorbed by the body B so that the temperature of the gas remains equal to t. As a result, the pressure increases according to Mariotte's law; it will be represented geometrically by the ordinates of a hyperbola KF and the corresponding abscissae will represent the volumes. Let us suppose that the compression is continued till the heat released by the compression of the gas and absorbed by the body *B* is exactly equal to the heat communicated by the source *A* to the gas, during its expansion in contact with it in the first part of the operation.*

Then let *AH* be the volume of the gas and *HK* the corresponding pressure. In this state the gas possesses the same absolute quantity of heat as at the start of the operation, when it occupied the volume *AB* under pressure *CB*. If then the body *B* is removed and the gas is further compressed inside an envelope which is impermeable to heat until the volume *AH* becomes the volume *AB*, its temperature increases all the time by the release of latent caloric made perceptible by the compression. At the same time the pressure increases, and when the volume is reduced to *AB* the temperature returns to *T* and the pressure to *BC*. Now the different states in which a given mass of gas can exist are characterized by the volume, pressure, temperature, and the absolute quantity of heat which it contains; if two of these four quantities are known, the other two are determined; thus, in the case under discussion, since the absolute quantity of heat and the volume are the same as they were at the start of the operation, it is certain that the temperature and the pressure will also be what they were then. Consequently the unknown law, of how the pressure varies when the volume of the gas is reduced inside its impermeable envelope, is represented by a curve *KC* which passes through point *C*, and whose abscissae and ordinates always represent volumes and pressures.

However, the reduction of the volume of the gas from *AC* to *AB* will have consumed a quantity of mechanical action which, by the same arguments which we have given above, will be represented by the two mixtilinear trapeziums *FGHK* and *KHBC*. If we subtract these two trapeziums from the first two *CBDE* and *EDFG* which represent the quantity of action developed during the expansion of the gas, the difference, which will be equal to the kind of curvilinear parallelogram *CEFK*, will represent the quantity of action developed by the cycle of operations just described, at the end of which the gas is in precisely its original state.

However, all the heat given up by the body *A* to the gas during its expansion in contact with it has flowed into the body *B* during the compression of the gas which took place in contact with that body.

* [Here "heat" should be taken to mean entropy. E. M.]

Here, therefore, mechanical force has been developed by the passage of caloric from a hot body to a cold body, and this passage has been accomplished without any contact between bodies at different temperatures.

The reverse operation is equally possible; thus let us take the same volume of gas AB at the temperature T and at pressure BC; let us enclose it in an envelope impermeable to heat, and dilate it so that its temperature falls gradually till it is equal to t; let us continue the expansion in the same envelope, but after having introduced the body B which has the same temperature; this provides the gas with the heat necessary to maintain its temperature and we continue the operation till the body B has given the gas the heat which it had received from it in the previous operation. Next let us remove the body B and compress the gas in an impermeable envelope till its temperature becomes once again equal to T. Then let us bring up the body A which has the same temperature, and continue to reduce the volume till all the heat taken from the body B has been given up to the body A. The gas then must have the same temperature and must possess the same absolute quantity of heat as at the beginning of the operation; whence it must occupy the same volume and must be at the same pressure.

Here the gas passes successively, but in the reverse order, through all the states of temperature and pressure through which it passed in the first series of operations; as a result the expansions have become compressions and vice versa; but they follow the same law. It follows that the quantities of action developed in the first case are absorbed in the second and vice versa, but they keep the same numerical values, because the elements of the integrals which compose them are the same.

Thus it can be seen that in causing heat to pass from a body maintained at a given temperature, using the method first described, a certain quantity of mechanical action is developed which is equal to that which must be used up to cause the same quantity of heat to pass from the cold body to the hot body by the reverse procedure just described.

A similar result can be arrived at through the vaporization of any liquid. Let us take such a liquid and put it in contact with the body A in a rigid envelope impermeable to heat; we suppose the temperature of the liquid to be equal to the temperature T of the body A. We mark on the axis of the abscissae AX (Fig. 2) a quantity AB equal to the volume of liquid and, on a line parallel to the axis of ordinates AY, a quantity BC equal to the vapor pressure of the liquid corresponding to the temperature T.

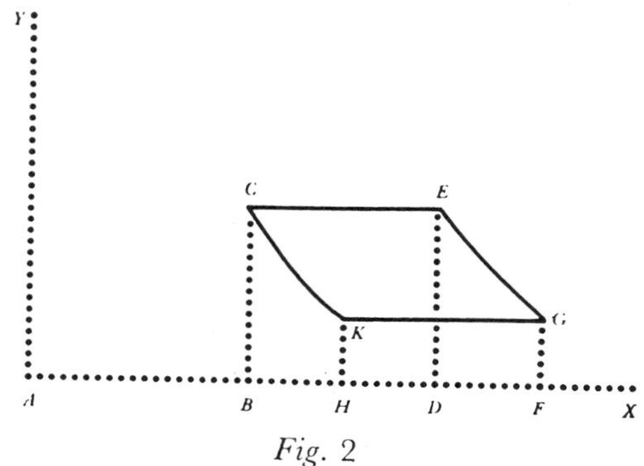

Fig. 2

If we increase the volume of the liquid, a part of it passes into the vapor state, and as the source of heat A provides the latent caloric necessary for its formation, the temperature remains constant and equal to T. If we mark on the axis of the abscissae quantities representing the successive volumes which the mixture of liquid and vapor occupy, and if the corresponding values of the pressure are taken for ordinates, then as the pressure remains constant the pressure curve is reduced here to a straight line parallel to the axis of the abscissae.

When a certain quantity of vapor has been formed and when the mixture of liquid and vapor occupies a volume AD, the body A is removed and the expansion continued. Then a further quantity of liquid passes to a gaseous state and a part of the perceptible caloric becomes latent, the temperature of the mixture falls as well as the pressure; let us suppose that the dilation is continued till the temperature, after falling gradually, becomes equal to the temperature t of the body B; let AF be the volume, FG the corresponding pressure. The law of the variation of pressure is given by some curve EG which passes through the points E and G.

During this first part of the operation a quantity of action will have been developed represented by the areas of the rectangle $BCED$ and the mixtilinear trapezium $EGFD$.

Let us now bring up the body B, put it in contact with the mixture of liquid and vapor and gradually reduce the volume; a part of the vapor passes to the liquid state and since the latent heat which it releases while condensing is absorbed by the body B as it is produced, the temperature remains constant equal to t. We continue to reduce the volume in this way till all the heat provided by the body A in the first part of the operation has been given to the body B.

Let AH be the volume then occupied by the mixture of vapor and liquid; the corresponding pressure is KH equal to GF; since the temperature remains equal to t during the reduction of the volume from AF to AH, the law of pressure between these two limits is represented by the line KG parallel to the axis of the abscissae.

Having arrived at this point, the mixture of vapor and liquid with which we are working, which occupies the volume AH under the pressure KH, and at a temperature t, possesses the same absolute quantity of heat that the liquid possessed at the start of the operation; if, therefore, the body B is removed and the condensation is continued in a vessel impermeable to heat, till the volume becomes again equal to AB, the same quantity of matter is occupying the same volume and possesses the same quantity of heat as at the beginning of the operation; its temperature and pressure must therefore be the same as at that stage; the temperature must also become again equal to CB; the law of pressures during this last part of the operation will therefore be given by a curve passing through the points K and C and the quantity of action absorbed during the reduction of the volume AF to AB is represented by the rectangle $FHKG$ and the mixtilinear trapezium $BCKH$.

If, therefore, the quantity of action developed during the expansion is subtracted from that which is absorbed during the compression, there remains the difference, the area of the mixtilinear parallelogram $CEGK$, which represents the quantity of action developed during the complete series of operations just described, at the end of which the liquid finds itself in its original state.

But it must be noted that all the caloric communicated by the body A has passed into the body B and that this transfer has been carried out without there having been any contact other than that between bodies at the same temperatures.

In the same way as for gases, it may be proved that by repeating the same operation in the reverse order, heat may be made to pass from the body B to the body A, but that this result can only be achieved by the absorption of a quantity of action equal to that developed by the passage of the same quantity of caloric from the body A to the body B.

From what has gone before, it follows that a quantity of mechanical action, and a quantity of heat which can pass from a hot body to a cold body, are quantities of the same nature, and that it is possible to replace the one by the other; in the same manner as in mechanics a body which is able to fall from a certain height and a mass moving with a certain velocity are quantities of the same order, which can be transformed one into the other by physical means.*

Hence it also follows that the quantity of action F developed by the passage of a certain quantity of heat C from a body maintained at a temperature t, by one of the procedures just outlined, is the same whatever the gas or liquid employed, and is the greatest which it is possible to attain. For suppose that by any process whatever the quantity of heat C were made to pass from the body A to the body B and that it was possible to produce a greater quantity of action F', we could use a part F to restore the quantity of heat C from the body B to the body A by one of the two methods just described; the *vis viva* F used for this purpose would, as we have seen, be equal to that developed by the passage of the same quantity of heat C from the body A to the body B; so by hypothesis it is smaller than F', so there would be produced a quantity of action $F'-F$, created out of nothing and without consumption of heat, an absurd result which would lead to the possibility of creating force or heat gratuitously and without limit. It seems to me that the impossibility of such a result can be accepted as a fundamental axiom of mechanics; no one has ever objected to Lagrange's demonstration of the principle of virtual velocities using pulleys, and this seems to me to depend on something similar.

In the same way it can be shown that no gas or vapor exists which, if used to transmit heat from a hot body to a cold one by the methods described, can develop a quantity of action greater than any other gas or vapor.

We therefore base our researches on the following principles: Caloric, passing from one body to another maintained at a smaller temperature, can give rise to the production of a certain quantity of mechanical action; there is a loss of *vis viva* whenever there is contact between bodies at different temperatures. The maximum effect is produced when the passage of the caloric from the hot to the cold body is effected by one of the methods just described. In

* [This extraordinary paragraph is an unambiguous statement of the First Law of Thermodynamics. It serves to emphasize the point made in the Introduction, that the caloric theory and the *vis viva* theory were not regarded as mutually exclusive. E. M.]

addition this is independent of the chemical nature of the liquid or gas employed, of its quantity or its pressure; so that the maximum quantity of action which the passage of a given quantity of heat from a cold to a hot body can develop is independent of the nature of the agents used.

§III

We will now translate analytically the various operations described in the previous paragraph; we will deduce the expression for the maximum quantity of action developed by the passage given quantity of heat from a body maintained at a certain temperature to a body maintained at a lower temperature, and we will arrive at some new relations between the volume, pressure, temperature and absolute heat or latent caloric of solids, liquids or gases.

Let us again consider the two bodies A and B and suppose that the temperature of the body B is smaller than the temperature t of the body A by an infinitely small amount dt. Let us first suppose that it is a gas which serves to transmit the caloric from the body A to the body B. Let V_0 be the volume of the gas at the pressure p_0 and at the temperature t_0; let p and v be the volume and the pressure of the same mass of the gas at the temperature t of the body A. Mariotte's law, combined with that of Gay-Lussac, establishes between these different quantities the relation

$$pv = \frac{p_0 v_0}{267 + t_0}(267 + t),$$

or simply

$$\frac{p_0 v_0}{267 + t_0} = R:$$

$$pv = R(267 + t).$$

The body A is put into contact with the gas. Let $me = v$, $ae = p$ (Fig. 3). If the gas is dilated by an infinitesimal amount $dv = eg$, the temperature remains constant because of the presence of the source of heat A; the pressure diminishes and becomes equal to the ordinate bg. Now the body A is removed and the gas dilated in an envelope impermeable to heat, by an infinitesimal amount gh, till the heat which has become latent lowers the temperature of the gas by an infinitesimal amount *dt* and thus brings it to the temperature $t-dt$ of the body *B*. As a result of this lowering of temperature

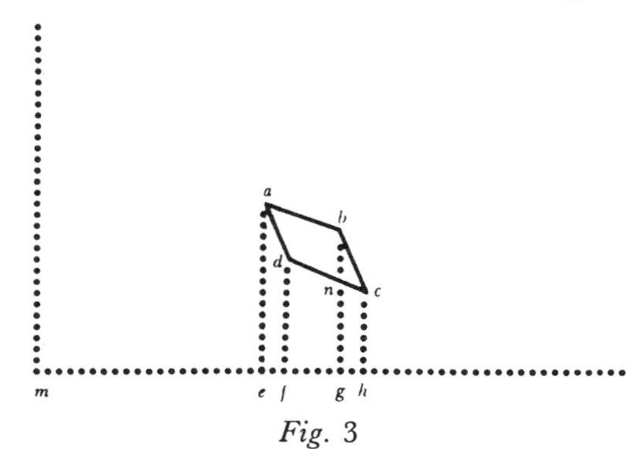

Fig. 3

the pressure falls more rapidly than in the first part of the operation and becomes *ch*. We now bring up the body *B* and reduce the volume *mh* by an infinitesimal amount *fh*, so calculated that during this compression the gas gives up to the body *B* all the heat which it drew from the body *A* during the first part of the operation. Let *fd* be the corresponding pressure; having done that we remove the body *B* and continue to compress the gas till it has regained the volume *me*. Then the pressure has become again equal to *ae* as demonstrated in the previous paragraph, and it can also be shown in the same way that the quadrilateral *abcd* measures the quantity of action produced by the transfer to the body *B* of the heat drawn from the body *A* during the dilation of the gas.

Now it is easy to show that this quadrilateral is a parallelogram; this is a result of the infinitely small values given to the variations of volume and pressure: for imagine that through each of the points in the plane on which the quadrilateral *abcd* is drawn, we erect perpendiculars to this plane, and that on each of them are marked two quantities T and Q as distances from the plane, the first equal to the temperature, the second to the absolute quantity of heat which the gas possesses when the volume and pressure have the values assigned to them by the abscissa v and ordinate p corresponding to each point.

The lines *ab* and *cd* belong to the projections of two equal-temperature curves on the temperature surface passing through two points infinitely close together; *ab* and *cd* are therefore parallel; *ad* and *bc* are also the projections of two curves for which $Q = \text{const.}$ on the surface $Q = f(p, v)$ which also passes through two infinitely close points, hence these two elements are also parallel; the quadrilateral *abcd* is therefore a parallelogram and it is easy to see that its area is found by multiplying the variation of volume during the contact of the gas with the body A or the body B, that is *eg*, or *fh* which is equal to it, by the difference *bn* of the pressure during these two operations and corresponding to the same value of the volume v. Now since *eg* and *fh* are the differentials of the volume, they are equal to dv; *bn* is obtained by differentiating the equation $pv = R(267 + t)$ holding v constant; therefore

$$bn = dp = R\frac{dt}{v}.$$

The quantity of action developed is therefore expressed as

$$R\frac{dt \cdot dv}{v}.$$

We still have to determine the quantity of heat needed to produce this effect; it is equal to that which the gas drew from the body A while its volume increased by dv keeping the same temperature; now since Q is the absolute quantity of heat which the gas possesses, it must be a certain function of p and v, taken as independent variables; the quantity of heat absorbed by the gas is therefore

$$dQ = \frac{dQ}{dv} \cdot dv + \frac{dQ}{dp} \cdot dp,$$

but the temperature remains constant during the variation of volume so

$$v \cdot dp + p \cdot dv = 0, \qquad \text{whence} \qquad dp = -\frac{p}{v} \cdot dv,$$

and hence

$$dQ = \left(\frac{dQ}{dv} - \frac{p}{v} \cdot \frac{dQ}{dp}\right) dv.$$

Dividing the effect produced by this value of dQ, we have

$$\frac{Rdt}{v\frac{dQ}{dv} - p\frac{dQ}{dp}}$$

as the expression for the maximum effect which a quantity of heat equal to unity can develop by passing from a body maintained at the temperature t to a body maintained at the temperature $t - dt$.

We have demonstrated that this quantity of action developed is independent of the agent which serves to transmit the heat; it is therefore the same for all gases, neither does it depend on the mass of the body employed; but there is nothing to show that it is independent of the temperature;

$$\left(v\frac{dQ}{dv} - p\frac{dQ}{dp}\right)$$

must therefore be equal to an unknown function of t which is the same for all gases.

Now because of the equation $pv = R(267+t)$, t is itself a function of the product pv, therefore we have the partial differential equation

$$v\frac{dQ}{dv} - p\frac{dQ}{dp} = F(p \cdot v);$$

this has for integral

$$Q = f(p \cdot v) - F(p \cdot v) \log_e p.$$

The generality of this formula is not at all changed by replacing these two arbitrary functions of the product pv by the functions B and C of the temperature, multiplied by the coefficient R; therefore we have

$$Q = R(B - C \log p).*$$

It is easy to verify that this value of Q satisfies all the required conditions; for we have

$$\frac{dQ}{dv} = R\left(\frac{dB}{dt} \cdot \frac{p}{R} - \log p \cdot \frac{dC}{dt} \cdot \frac{p}{R}\right)$$

$$\frac{dQ}{dp} = R\left(\frac{dB}{dt} \cdot \frac{v}{R} - \log p \cdot \frac{dC}{dt} \cdot \frac{v}{R} - C\frac{1}{p}\right);$$

from this is found

$$v\frac{dQ}{dv} - p\frac{dQ}{dp} = CR$$

and consequently

$$\frac{Rdt}{v\dfrac{dQ}{dv} - p\dfrac{dQ}{dp}} = \frac{dt}{C}.$$

* [This corresponds to an integrated form of the equation for a reversible change with a perfect gas:

$$T \cdot dS = C_p \cdot dT - \frac{RT}{J} \cdot \frac{dp}{p}$$

where R is expressed in mechanical units and J is the mechanical equivalent of heat. The function C is in fact equal to T/J. E. M.]

The function C which multiplies the logarithm of the pressure in the value of Q is of great importance, as can be seen; it is independent of the nature of the gas and is a function of the temperature only; it is essentially positive and serves as a measure of the maximum quantity of action which heat can develop.

We have seen that of the four quantities Q, t, p and v, if two are known the other two are determined; they must therefore be connected by two equations; one of them,

$$pv = R(267+t),$$

results from the laws of Mariotte and Gay-Lussac combined. The equation

$$Q = R(B-C\log p),$$

which we deduce from our theory, is the second. However, the numerical calculation of the changes which gases undergo when the volume and pressure are varied arbitrarily, demands a knowledge of the functions B and C.

We will see later that approximate values of the function C can be obtained over a considerable range of temperatures; further, having been obtained for one gas it will be the same for all. As to the function B, it can vary from one gas to another; however, it is probable that it is the same for all simple gases: at least, that seems to follow from experiment, from the fact that they have the same heat capacities.

Let us return to the equation

$$Q = R(B-C\log p).$$

Let us compress a gas occupying volume v at pressure p till the volume becomes v', and let it be cooled till the temperature returns to the same point. Let p' be the new value of the pressure; let Q' be the new value of Q; we have

$$Q-Q' = RC\log\frac{p'}{p} = RC\log\frac{v}{v'}.$$

Since the function C is the same for all gases, it is seen that *equal volumes of all elastic fluids, taken at the same temperature and under the same pressure, when compressed or expanded by the same fraction of their volumes, set free or absorb the same absolute quantity of heat.* This is the law deduced by Dulong by direct experiment.

In addition this equation shows that *when a gas varies its volume without changing its temperature, the quantities of heat absorbed or set free are in arithmetic progression if the increments or reductions of volume are in geometrical progression.* This result has also been enunciated by Carnot in the work cited.

The equation

$$Q - Q' = RC \log \left(\frac{v}{v'}\right)$$

expresses a more general law; it takes account of all the circumstances which can influence the phenomenon, such as the pressure, the volume and the temperature.

For since

$$R = \frac{p_0 v_0}{267 + t_0} = \frac{pv}{267 + t},$$

we have

$$Q - Q' = \frac{pv}{267 + t} C \log \frac{v}{v'}.$$

This equation shows the influence of pressure; it shows that *equal volumes of all gases, taken at the same temperature, if compressed or expanded by a given fraction of their volumes, set free or absorb quantities of heat proportional to the pressure.*

This result explains how the sudden entry of air into the receiver of the pneumatic pump does not release a perceptible quantity of heat. The vacuum of the pneumatic pump is nothing more than a volume v of gas whose pressure p is very small; if atmospheric air is allowed to enter, its pressure p becomes suddenly equal to that of the atmosphere, p', its volume is reduced to v' and the expression for the heat released is

$$C \frac{pv}{267 + t} \log \frac{v}{v'} = C \frac{pv}{267 + t} \log \frac{p'}{p}.$$

The heat released by the entry of the atmospheric air into the vacuum is therefore given by the expression when p is made very small; then $\log \frac{p'}{p}$ is very big, but the product of p with $\log \frac{p'}{p}$ is very small; for

$$p \log \frac{p'}{p} = p \log p' - p \log p = p(\log p' - \log p),$$

a quantity which converges to zero when p diminishes.

The heat released is therefore smaller the lower the pressure in the receiver, and it falls to zero when the vacuum is perfect.

We will add that the equation

$$Q = R(B - C \log p)$$

gives the law of specific heats at constant pressure and constant volume.* The first is expressed as

$$R\left(\frac{dB}{dt} - \frac{dC}{dt}\log p\right);$$

the second as

$$R\left(\frac{dB}{dt} - \frac{dC}{dt}\log p - C\frac{1}{p}\frac{dp}{dt}\right),$$

which is equal to

$$R\left(\frac{dB}{dt} - \frac{dC}{dt}\log p - \frac{C}{267+t}\right).$$

The first is obtained by differentiating Q with respect to t keeping p constant; the second keeping v constant. If equal volumes of different gases are taken at the same temperature and at the same pressure, the quantity R is the same for all and consequently it is seen that the specific heat at constant pressure exceeds the specific heat at constant volume by an amount which is the same for all gases, equal to

$$\frac{R}{267+t}\cdot C.$$

[*Editor's Note:* Material has been omitted at this point.]

* [These are almost the only ones of Clapeyron's equations for observable quantities which are actually wrong. They imply that the specific heats of a perfect gas depend on pressure, which was consistent with some of Delaroche and Bérard's data. But he does not follow up these implications. E. M.]

3

Reprinted from pp. 100–106 of *Mathematical and Physical Papers of William Thomson,* Vol. 1, Cambridge University Press, 1882, 571 pp.

ON AN ABSOLUTE THERMOMETRIC SCALE FOUNDED ON CARNOT'S THEORY OF THE MOTIVE POWER OF HEAT,* AND CALCULATED FROM REGNAULT'S OBSERVATIONS†

W. Thomson

THE determination of temperature has long been recognized as a problem of the greatest importance in physical science. It has accordingly been made a subject of most careful attention, and, especially in late years, of very elaborate and refined experimental researches‡; and we are thus at present in possession of as complete a practical solution of the problem as can be desired, even for the most accurate investigations. The theory of thermometry is however as yet far from being in so satisfactory a state. The principle to be followed in constructing a thermometric scale might at first sight seem to be obvious, as it might appear that a perfect thermometer would indicate equal additions of heat, as corresponding to equal elevations of temperature, estimated by the numbered divisions of its scale. It is however now recognized (from the variations in the specific heats of bodies) as an experimentally demonstrated fact that thermometry under this condition is impossible,

* Published in 1824 in a work entitled *Réflexions sur la Puissance Motrice du Feu*, by M. S. Carnot. Having never met with the original work, it is only through a paper by M. Clapeyron, on the same subject, published in the *Journal de l'École Polytechnique*, Vol. XIV. 1834, and translated in the first volume of Taylor's *Scientific Memoirs*, that the Author has become acquainted with Carnot's Theory.—W. T. [Note of Nov. 5th, 1881. A few months later through the kindness of my late colleague Prof. Lewis Gordon, I received a copy of Carnot's original work and was thus enabled to give to the Royal Society of Edinburgh my "Account of Carnot's theory" which is reprinted as Art. XLI. below. The original work has since been republished, with a biographical notice, Paris, 1878.]

† An account of the first part of a series of researches undertaken by M. Regnault by order of the French Government, for ascertaining the various physical data of importance in the Theory of the Steam Engine, is just published in the *Mémoires de l'Institut*, of which it constitutes the twenty-first volume (1847). The second part of the researches has not yet been published. [Note of Nov. 5, 1881. The continuation of these researches has now been published: thus we have for the whole series, Vol. I. in 1847; Vol. II. in 1862; and Vol. III. in 1870.]

‡ A very important section of Regnault's work is devoted to this object.

and we are left without any principle on which to found an absolute thermometric scale.

Next in importance to the primary establishment of an absolute scale, independently of the properties of any particular kind of matter, is the fixing upon an arbitrary system of thermometry, according to which results of observations made by different experimenters, in various positions and circumstances, may be exactly compared. This object is very fully attained by means of thermometers constructed and graduated according to the clearly defined methods adopted by the best instrument-makers of the present day, when the rigorous experimental processes which have been indicated, especially by Regnault, for interpreting their indications in a comparable way, are followed. The particular kind of thermometer which is least liable to uncertain variations of any kind is that founded on the expansion of air, and this is therefore generally adopted as the standard for the comparison of thermometers of all constructions. Hence the scale which is at present employed for estimating temperature is that of the air-thermometer; and in accurate researches care is always taken to reduce to this scale the indications of the instrument actually used, whatever may be its specific construction and graduation.

The principle according to which the scale of the air-thermometer is graduated, is simply that equal absolute expansions of the mass of air or gas in the instrument, under a constant pressure, shall indicate equal differences of the numbers on the scale; the length of a "degree" being determined by allowing a given number for the interval between the freezing- and the boiling-points. Now it is found by Regnault that various thermometers, constructed with air under different pressures, or with different gases, give indications which coincide so closely, that, unless when certain gases, such as sulphurous acid, which approach the physical condition of vapours at saturation, are made use of, the variations are inappreciable*. This remarkable circumstance enhances very much the practical value of the air-thermometer; but still a

* Regnault, *Relation des Expériences*, &c., Fourth Memoir, First Part. The differences, it is remarked by Regnault, would be much more sensible if the graduation were effected on the supposition that the coefficients of expansion of the different gases are equal, instead of being founded on the principle laid down in the text, according to which the freezing- and boiling-points are experimentally determined for each thermometer.

rigorous standard can only be defined by fixing upon a certain gas at a determinate pressure, as the thermometric substance. Although we have thus a strict principle for constructing a *definite* system for the estimation of temperature, yet as reference is essentially made to a specific body as the standard thermometric substance, we cannot consider that we have arrived at an *absolute* scale, and we can only regard, in strictness, the scale actually adopted as *an arbitrary series of numbered points of reference sufficiently close for the requirements of practical thermometry.*

In the present state of physical science, therefore, a question of extreme interest arises: *Is there any principle on which an absolute thermometric scale can be founded?* It appears to me that Carnot's theory of the motive power of heat enables us to give an affirmative answer.

The relation between motive power and heat, as established by Carnot, is such that *quantities of heat*, and *intervals of temperature*, are involved as the sole elements in the expression for the amount of mechanical effect to be obtained through the agency of heat; and since we have, independently, a definite system for the measurement of quantities of heat, we are thus furnished with a measure for intervals according to which absolute differences of temperature may be estimated. To make this intelligible, a few words in explanation of Carnot's theory must be given; but for a full account of this most valuable contribution to physical science, the reader is referred to either of the works mentioned above (the original treatise by Carnot, and Clapeyron's paper on the same subject.

In the present state of science no operation is known by which heat can be absorbed, without either elevating the temperature of matter, or becoming latent and producing some alteration in the physical condition of the body into which it is absorbed; and the conversion of heat (or *caloric*) into mechanical effect is probably impossible*, certainly undiscovered. In actual engines for ob-

* This opinion seems to be nearly universally held by those who have written on the subject. A contrary opinion however has been advocated by Mr Joule of Manchester; some very remarkable discoveries which he has made with reference to the *generation* of heat by the friction of fluids in motion, and some known experiments with magneto-electric machines, seeming to indicate an actual conversion of mechanical effect into caloric. No experiment however is adduced in which the converse operation is exhibited; but it must be confessed that as yet much is involved in mystery with reference to these fundamental questions of natural philosophy.

taining mechanical effect through the agency of heat, we must consequently look for the source of power, not in any absorption and conversion, but merely in a transmission of heat. Now Carnot, starting from universally acknowledged physical principles, demonstrates that it is by the *letting down* of heat from a hot body to a cold body, through the medium of an engine (a steam-engine, or an air-engine for instance), that mechanical effect is to be obtained; and conversely, he proves that the same amount of heat may, by the expenditure of an equal amount of labouring force, be *raised* from the cold to the hot body (the engine being in this case *worked backwards*); just as mechanical effect may be obtained by the descent of water let down by a water-wheel, and by spending labouring force in turning the wheel backwards, or in working a pump, water may be elevated to a higher level. The amount of mechanical effect to be obtained by the transmission of a given quantity of heat, through the medium of any kind of engine in which the economy is perfect, will depend, as Carnot demonstrates, not on the specific nature of the substance employed as the medium of transmission of heat in the engine, but solely on the interval between the temperature of the two bodies between which the heat is transferred.

Carnot examines in detail the ideal construction of an air-engine and of a steam-engine, in which, besides the condition of perfect economy being satisfied, the machine is so arranged, that at the close of a complete operation the substance (air in one case and water in the other) employed is restored to precisely the same physical condition as at the commencement. He thus shews on what elements, capable of experimental determination, either with reference to air, or with reference to a liquid and its vapour, the absolute amount of mechanical effect due to the transmission of a unit of heat from a hot body to a cold body, through any given interval of the thermometric scale, may be ascertained. In M. Clapeyron's paper various experimental data, confessedly very imperfect, are brought forward, and the amounts of mechanical effect due to a unit of heat descending a degree of the air-thermometer, in various parts of the scale, are calculated from them, according to Carnot's expressions. The results so obtained indicate very decidedly, that what we may with much propriety call *the value of a degree* (estimated by the mechanical effect to be obtained from the descent of a unit of heat through it) of the air-thermometer depends on the part of the scale in which it is taken, being less for high than for low temperatures*.

The characteristic property of the scale which I now propose is, that all degrees have the same value; that is, that a unit of heat descending from a body A at the temperature T^0 of this scale, to a body B at the temperature $(T-1)^0$, would give out the same mechanical effect, whatever be the number T. This may justly be termed an absolute scale, since its characteristic is quite independent of the physical properties of any specific substance.

To compare this scale with that of the air-thermometer, the *values* (according to the principle of estimation stated above) of degrees of the air-thermometer must be known. Now an expression, obtained by Carnot from the consideration of his ideal steam-engine, enables us to calculate these values, when the latent heat of a given volume and the pressure of saturated vapour at any temperature are experimentally determined. The determination of these elements is the principal object of Regnault's great work, already referred to, but at present his researches are not complete. In the first part, which alone has been as yet published, the latent heats of a given *weight*, and the pressures of saturated vapour, at all temperatures between 0^0 and 230^0 (Cent. of the air-thermometer), have been ascertained; but it would be necessary in addition to know the densities of saturated vapour at different temperatures, to enable us to determine the latent heat of a given volume at any temperature. M. Regnault announces his intention of instituting researches for this object; but till the results are made known, we have no way of completing the data necessary for the present problem, except by estimating the density of saturated vapour at any temperature (the corresponding pressure being known by Regnault's researches already published) according to the approximate laws of compressibility and expansion (the laws

* This is what we might anticipate, when we reflect that infinite cold must correspond to a finite number of degrees of the air-thermometer below zero; since, if we push the strict principle of graduation, stated above, sufficiently far, we should arrive at a point corresponding to the volume of air being reduced to nothing, which would be marked as -273^0 of the scale ($-100/\cdot366$, if $\cdot366$ be the coefficient of expansion); and therefore -273^0 of the air-thermometer is a point which cannot be reached at any finite temperature, however low.

of Mariotte and Gay-Lussac, or Boyle and Dalton). Within the limits of natural temperature in ordinary climates, the density of saturated vapour is actually found by Regnault (*Études Hygrométriques* in the *Annales de Chimie*) to verify very closely these laws; and we have reason to believe from experiments which have been made by Gay-Lussac and others, that as high as the temperature 100° there can be no considerable deviation; but our estimate of the density of saturated vapour, founded on these laws, may be very erroneous at such high temperatures as 230°. Hence a completely satisfactory calculation of the proposed scale cannot be made till after the additional experimental data shall have been obtained; but with the data which we actually possess, we may make an approximate comparison of the new scale with that of the air-thermometer, which at least between 0° and 100° will be tolerably satisfactory.

The labour of performing the necessary calculations for effecting a comparison of the proposed scale with that of the air-thermometer, between the limits 0° and 230° of the latter, has been kindly undertaken by Mr William Steele, lately of Glasgow College, now of St Peter's College, Cambridge. His results in tabulated forms were laid before the Society, with a diagram, in which the comparison between the two scales is represented graphically. In the first table*, the amounts of mechanical effect due to the descent of a unit of heat through the successive degrees of the air-thermometer are exhibited. The unit of heat adopted is the quantity necessary to elevate the temperature of a kilogramme of water from 0° to 1° of the air-thermometer; and the unit of mechanical effect is a metre-kilogramme; that is, a kilogramme raised a metre high.

In the second table, the temperatures according to the proposed scale, which correspond to the different degrees of the air-thermometer from 0° to 230°, are exhibited. [The arbitrary points which coincide on the two scales are 0° and 100°].

Note.—If we add together the first hundred numbers given in the first table, we find 135·7 for the amount of work due to a unit of heat descending from a body A at 100° to B at 0°. Now 79 such units of heat would, according to Dr Black (his result being

* [Note of Nov. 4, 1881. This table (reduced from metres to feet) was repeated in my "Account of Carnot's Theory of the Motive power of Heat," republished as Article XLI. below, in § 38 of which it will be found.]

very slightly corrected by Regnault), melt a kilogramme of ice. Hence if the heat necessary to melt a pound of ice be now taken as unity, and if a *metre-pound* be taken as the unit of mechanical effect, the amount of work to be obtained by the descent of a unit of heat from 100° to 0° is 79 × 135·7, or 10,700 nearly. This is the same as 35,100 foot pounds, which is a little more than the work of a one-horse-power engine (33,000 foot pounds) in a minute; and consequently, if we had a steam-engine working with perfect economy at one-horse-power, the boiler being at the temperature 100°, and the condenser kept at 0° by a constant supply of ice, rather less than a pound of ice would be melted in a minute.

[Note of Nov. 4, 1881. This paper was wholly founded on Carnot's uncorrected theory, according to which the quantity of heat taken in in the hot part of the engine, (the boiler of the steam engine for instance), was supposed to be equal to that abstracted from the cold part (the condenser of the steam engine), in a complete period of the regular action of the engine, when every varying temperature, in every part of the apparatus, has become strictly periodic. The reconciliation of Carnot's theory with what is now known to be the true nature of heat is fully discussed in Article XLVIII. below; and in §§ 24—41 of that article, are shewn in detail the consequently required corrections of the thermodynamic estimates of the present article. These corrections however do not in any way affect the absolute scale for thermometry which forms the subject of the present article. Its relation to the practically more convenient scale (agreeing with air thermometers nearly enough for most purposes, throughout the range from the lowest temperatures hitherto measured, to the highest that can exist so far as we know) which I gave subsequently, Dynamical Theory of Heat (Art. XLVIII. below), Part VI., §§ 99, 100; *Trans. R. S. E.*, May, 1854: and Article 'Heat,' §§ 35—38, 47—67, *Encyclopædia Britannica*, is shewn in the following formula:

$$\theta = 100 \frac{\log t - \log 273}{\log 373 - \log 273},$$

where θ and t are the reckonings of one and the same temperature, according to my first and according to my second thermodynamic absolute scale.]

4

Reprinted from pp. 113–140 of *Mathematical and Physical Papers of William Thomson,* Vol. 1, Cambridge University Press, 1882, 571 pp.

AN ACCOUNT OF CARNOT'S THEORY OF THE MOTIVE POWER OF HEAT*; WITH NUMERICAL RESULTS DEDUCED FROM REGNAULT'S EXPERIMENTS ON STEAM†

W. Thomson

(Read January 2, 1849.)

1. THE presence of heat may be recognised in every natural object; and there is scarcely an operation in nature which is not more or less affected by its all-pervading influence. An evolution and subsequent absorption of heat generally give rise to a variety of effects; among which may be enumerated, chemical combinations or decompositions; the fusion of solid substances; the vaporisation of solids or liquids; alterations in the dimensions of bodies, or in the statical pressure by which their dimensions may be modified; mechanical resistance overcome; electrical currents generated. In many of the actual phenomena of nature, several or all of these effects are produced together; and their complication will, if we attempt to trace the agency of heat in producing any individual effect, give rise to much perplexity. It will, therefore, be desirable, in laying the foundation of a physical

* Published in 1824, in a work entitled, "Réflexions sur la Puissance Motrice du Feu, et sur les Machines Propres à Déveloper cette Puissance. Par S. Carnot." An account of Carnot's Theory is also published in the *Journal de l'Ecole Polytechnique*, Vol. XIV., 1834, in a paper by Mons. Clapeyron. [Note of Nov. 5, 1881. The original work has now been republished, with a biographical notice, Paris, 1878.]

† An account of the first part of a series of researches undertaken by Mons. Regnault, by order of the late French Government, for ascertaining the various physical data of importance in the theory of the steam-engine, has been recently published (under the title, "Relation des Expériences," &c.) in the *Mémoires de l'Institut*, of which it constitutes the twenty-first volume (1847). The second part of these researches has not yet been published. [Note of Nov. 5, 1881. The continuation of these researches has now been published; thus we have for the whole series, Vol. I. in 1847; Vol. II. in 1862; and Vol. III. in 1870.]

theory of any of the effects of heat, to discover or to imagine phenomena free from all such complication, and depending on a definite thermal agency; in which the relation between the cause and effect, traced through the medium of certain simple operations, may be clearly appreciated. Thus it is that Carnot, in accordance with the strictest principles of philosophy, enters upon the investigation of the theory of the motive power of heat.

2. The sole effect to be contemplated in investigating the motive power of heat is *resistance overcome*, or, as it is frequently called, "*work performed*," or "*mechanical effect*." The questions to be resolved by a complete theory of the subject are the following:

(1) What is the precise nature of the thermal agency by means of which *mechanical effect* is to be produced, without effects of any other kind?

(2) How may the amount of this thermal agency necessary for performing a given quantity of work be estimated?

3. In the following paper I shall commence by giving a short abstract of the reasoning by which Carnot is led to an answer to the first of these questions; I shall then explain the investigation by which, in accordance with his theory, the experimental elements necessary for answering the second question are indicated; and, in conclusion, I shall state the *data* supplied by Regnault's recent observations on steam, and apply them to obtain, as approximately as the present state of experimental science enables us to do, a complete solution of the question.

I. On the nature of Thermal agency, considered as a motive power.

4. There are [at present known] two, and only two, distinct ways in which mechanical effect can be obtained from heat. One of these is by means of the alterations of volume which bodies may experience through the action of heat; the other is through the medium of electric agency. Seebeck's discovery of thermo-electric currents enables us at present to conceive of an electro-magnetic engine supplied from a thermal origin, being used as a motive power: but this discovery was not made until 1821, and the subject of thermo-electricity can only have been

generally known in a few isolated facts, with reference to the electrical effects of heat upon certain crystals, at the time when Carnot wrote. He makes no allusion to it, but confines himself to the method for rendering thermal agency available as a source of mechanical effect, by means of the expansions and contractions of bodies.

5. A body expanding or contracting under the action of force, may, in general, either produce mechanical effect by overcoming resistance, or receive mechanical effect by yielding to the action of force. The amount of mechanical effect thus developed will depend not only on the calorific agency concerned, but also on the alteration in the physical condition of the body. Hence, after allowing the volume and temperature of the body to change, we must restore it to its original temperature and volume; and then we may estimate the aggregate amount of mechanical effect developed as due solely to the thermal origin.

6. Now the ordinarily-received, and almost universally-acknowledged, principles with reference to "quantities of caloric" and "latent heat," lead us to conceive that, at the end of a cycle of operations, when a body is left in precisely its primitive physical condition, if it has absorbed any heat during one part of the operations, it must have given out again exactly the same amount during the remainder of the cycle. The truth of this principle is considered as axiomatic by Carnot, who admits it as the foundation of his theory; and expresses himself in the following terms regarding it, in a note on one of the passages of his treatise*.

"In our demonstrations we tacitly assume that after a body has experienced a certain number of transformations, if it be brought identically to its primitive physical state as to density, temperature, and molecular constitution, it must contain the same quantity of heat as that which it initially possessed; or, in other words, we suppose that the quantities of heat lost by the body under one set of operations are precisely compensated by those which are absorbed in the others. This fact has never been doubted; it has at first been admitted without reflection, and afterwards verified, in many cases, by calorimetrical experiments. To deny it would be to overturn the whole theory

* Carnot, p. 37.

of heat, in which it is the fundamental principle. It must be admitted, however, that the chief foundations on which the theory of heat rests, would require a most attentive examination. Several experimental facts appear nearly inexplicable in the actual state of this theory."

7. Since the time when Carnot thus expressed himself, the necessity of a most careful examination of the entire experimental basis of the theory of heat has become more and more urgent. Especially all those assumptions depending on the idea that heat is a *substance*, invariable in quantity; not convertible into any other element, and incapable of being *generated* by any physical agency; in fact the acknowledged principles of latent heat; would require to be tested by a most searching investigation before they ought to be admitted, as they usually have been, by almost every one who has been engaged on the subject, whether in combining the results of experimental research, or in general theoretical investigations.

8. The extremely important discoveries recently made by Mr Joule of Manchester, that heat is evolved in every part of a closed electric conductor, moving in the neighbourhood of a magnet*, and that heat is *generated* by the friction of fluids in

* The *evolution* of heat in a fixed conductor, through which a galvanic current is sent from any source whatever, has long been known to the scientific world; but it was pointed out by Mr Joule that we cannot infer from any previously-published experimental researches, the actual *generation* of heat when the current originates in electro-magnetic induction; since the question occurs, *is the heat which is evolved in one part of the closed conductor merely transferred from those parts which are subject to the inducing influence?* Mr Joule, after a most careful experimental investigation with reference to this question, finds that it must be answered in the negative.—(See a paper "*On the Calorific Effects of Magneto-Electricity, and on the Mechanical Value of Heat;* by J. P. Joule, Esq." Read before the British Association at Cork in 1843, and subsequently communicated by the Author to the *Philosophical Magazine*, Vol. XXIII., pp. 263, 347, 435.)

Before we can finally conclude that heat is absolutely generated in such operations, it would be necessary to prove that the inducing magnet does not become lower in temperature, and thus compensate for the heat evolved in the conductor. I am not aware that any examination with reference to the truth of this conjecture has been instituted; but, in the case where the inducing body is a pure electro-magnet (without any iron), the experiments actually performed by Mr Joule render the conclusion probable that the heat evolved in the wire of the electro-magnet is not affected by the inductive action, otherwise than through the reflected influence which increases the strength of its own current.

motion, seem to overturn the opinion commonly held that heat cannot be *generated*, but only produced from a source, where it has previously existed either in a sensible or in a latent condition.

In the present state of science, however, no operation is known by which heat can be absorbed into a body without either elevating its temperature, or becoming latent, and producing some alteration in its physical condition; and the fundamental axiom adopted by Carnot may be considered as still the most probable basis for an investigation of the motive power of heat; although this, and with it every other branch of the theory of heat, may ultimately require to be reconstructed upon another foundation, when our experimental data are more complete. On this understanding, and to avoid a repetition of doubts, I shall refer to Carnot's fundamental principle, in all that follows, as if its truth were thoroughly established.

9. We are now led to the conclusion that the origin of motive power, developed by the alternate expansions and contractions of a body, must be found in the agency of heat entering the body and leaving it; since there cannot, at the end of a complete cycle, when the body is restored to its primitive physical condition, have been any absolute absorption of heat, and consequently no conversion of heat, or caloric, into mechanical effect; and it remains for us to trace the precise nature of the circumstances under which heat must enter the body, and afterwards leave it, so that mechanical effect may be produced. As an example, we may consider that machine for obtaining motive power from heat with which we are most familiar—the steam-engine.

10. Here, we observe, that heat enters the machine from the furnace, through the sides of the boiler, and that heat is continually abstracted by the water employed for keeping the condenser cool. According to Carnot's fundamental principle, the quantity of heat thus discharged, during a complete revolution (or double stroke) of the engine must be precisely equal to that which enters the water of the boiler*; provided the total mass

* So generally is Carnot's principle tacitly admitted as an axiom, that its application in this case has never, so far as I am aware, been questioned by practical engineers.

of water and steam be invariable, and be restored to its primitive physical condition (which will be the case rigorously, if the condenser be kept cool by the external application of cold water, instead of by injection, as is more usual in practice), and if the condensed water be restored to the boiler at the end of each complete revolution. Thus, we perceive, that a certain quantity of heat is *let down* from a hot body, the metal of the boiler, to another body at a lower temperature, the metal of the condenser; and that there results from this transference of heat a certain development of mechanical effect.

11. If we examine any other case in which mechanical effect is obtained from a thermal origin, by means of the alternate expansions and contractions of any substance whatever, instead of the water of a steam-engine, we find that a similar transference of heat is effected, and we may therefore answer the first question proposed, in the following manner:—

The thermal agency by which mechanical effect may be obtained, is the transference of heat from one body to another at a lower temperature.

II. On the measurement of Thermal Agency, considered with reference to its equivalent of mechanical effect.

12. A *perfect* thermo-dynamic engine of any kind, is a machine by means of which the greatest possible amount of mechanical effect can be obtained from a given thermal agency; and, therefore, if in any manner we can construct or imagine a perfect engine which may be applied for the transference of a given quantity of heat from a body at any given temperature, to another body, at a lower given temperature, and if we can evaluate the mechanical effect thus obtained, we shall be able to answer the question at present under consideration, and so to complete the theory of the motive power of heat. But whatever kind of engine we may consider with this view, it will be necessary for us to prove that it is a perfect engine; since the transference of the heat from one body to the other may be wholly, or partially, effected by conduction through a solid*, without the

* When "thermal agency" is thus spent in conducting heat through a solid, what becomes of the mechanical effect which it might produce? Nothing can be lost in the operations of nature—no energy can be destroyed. What effect then is

development of mechanical effect; and, consequently, engines may be constructed in which the whole, or any portion of the thermal agency is wasted. Hence it is of primary importance to discover the criterion of a perfect engine. This has been done by Carnot, who proves the following proposition :—

13. *A perfect thermo-dynamic engine is such that, whatever amount of mechanical effect it can derive from a certain thermal agency; if an equal amount be spent in working it backwards, an equal reverse thermal effect will be produced* *.

14. This proposition will be made clearer by the applications of it which are given below (§ 29), in the cases of the air-engine and the steam-engine, than it could be by any general explanation; and it will also appear, from the nature of the operations described in those cases, and the principles of Carnot's reasoning, that a perfect engine may be constructed with any substance of an indestructible texture as the alternately expanding and contracting medium. Thus we might conceive thermo-dynamic engines founded upon the expansions and contractions of a perfectly elastic solid, or of a liquid; or upon the alterations of volume experienced by substances, in passing from the liquid to

produced in place of the mechanical effect which is lost? A perfect theory of heat imperatively demands an answer to this question; yet no answer can be given in the present state of science. A few years ago, a similar confession must have been made with reference to the mechanical effect lost in a fluid set in motion in the interior of a rigid closed vessel, and allowed to come to rest by its own internal friction; but in this case, the foundation of a solution of the difficulty has been actually found, in Mr Joule's discovery of the generation of heat, by the internal friction of a fluid in motion. Encouraged by this example, we may hope that the very perplexing question in the theory of heat, by which we are at present arrested, will, before long, be cleared up. [Note of Sep. 1881. The Theory of the Dissipation of Energy (Article LVIII., below) completely answers this question and removes the difficulty.]

It might appear, that the difficulty would be entirely avoided, by abandoning Carnot's fundamental axiom; a view which is strongly urged by Mr Joule (at the conclusion of his paper "On the Changes of Temperature produced by the Rarefaction and Condensation of Air." *Phil. Mag.*, May 1845, Vol. XXVI.) If we do so, however, we meet with innumerable other difficulties—insuperable without farther experimental investigation, and an entire reconstruction of the theory of heat from its foundation. It is in reality to experiment that we must look—either for a verification of Carnot's axiom, and an explanation of the difficulty we have been considering; or for an entirely new basis of the Theory of Heat.

* For a demonstration, see § 29, below.

the solid state*, each of which being perfect, would produce the same amount of mechanical effect from a given thermal agency; but there are two cases which Carnot has selected as most worthy of minute attention, because of their peculiar appropriateness for illustrating the general principles of his theory, no less than on account of their very great practical importance; the steam-engine, in which the substance employed as the transferring medium is water, alternately in the liquid state, and in the state of vapour; and the air-engine, in which the transference is effected by means of the alternate expansions and contractions of a medium, always in the gaseous state. The details of an actually practicable engine of either kind are not contemplated by Carnot, in his general theoretical reasonings, but he confines himself to the ideal construction, in the simplest possible way in each case, of an engine in which the economy is perfect. He thus determines the degree of perfectibility which cannot be surpassed; and, by describing a conceivable method of attaining to this perfection by an air-engine or a steam-engine, he points out the proper objects to be kept in view in the practical construction and working of such machines. I now proceed to give an outline of these investigations.

CARNOT'S *Theory of the Steam-Engine.*

15. Let CDF_2E_2 be a cylinder, of which the curved surface is perfectly impermeable to heat, with a piston also impermeable to heat, fitted in it; while the fixed bottom CD, itself with no capacity for heat, is possessed of perfect conducting power. Let K be an impermeable stand, such that when the cylinder is placed upon it, the contents below the piston can neither gain nor lose heat. Let A and B be two bodies permanently retained at constant temperatures, S^0 and T^0, respectively, of which the former is higher than the latter. Let the cylinder, placed on the impermeable stand, K, be partially filled with water, at the temperature S, of the body A, and (there being no air below it) let the piston be placed in a position EF, near the surface of

* A case minutely examined in another paper, to be laid before the Society at the present meeting. "Theoretical considerations on the Effect of Pressure in lowering the Freezing Point of Water," by Prof. James Thomson. [Appended at the end of the present Article by his permission Nov. 5, 1881.]

the water. The pressure of the vapour above the water will tend to push up the piston, and must be resisted by a force applied to the piston*, till the commencement of the operations, which are conducted in the following manner.

(1) The cylinder being placed on the body A, so that the water and vapour may be retained at the temperature S, *let the*

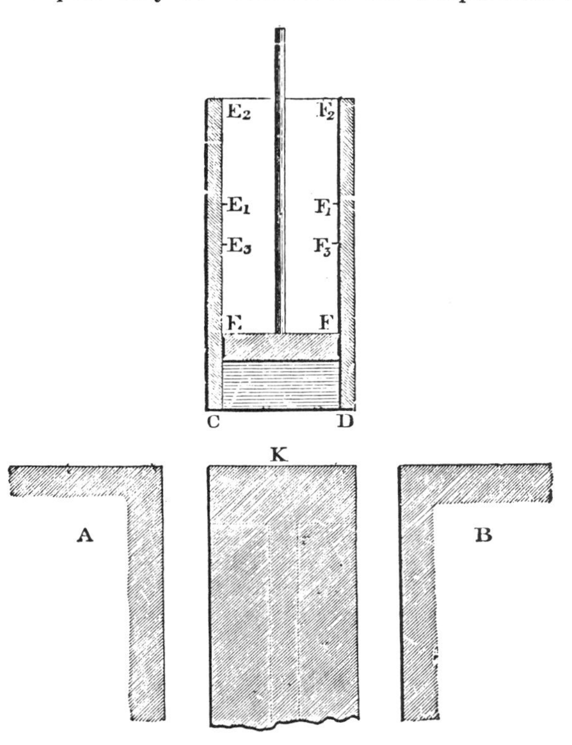

piston rise any convenient height EE_1, *to a position* E_1F_1, *performing work by the pressure of the vapour below it during its ascent.*

[During this operation a certain quantity, H, of heat, the amount of latent heat in the fresh vapour which is formed, is abstracted from the body A.]

* In all that follows, the pressure of the atmosphere on the upper side of the piston will be included in the applied forces, which, in the successive operations described, are sometimes overcome by the upward motion, and sometimes yielded to in the motion downwards. It will be unnecessary, in reckoning at the end of a cycle of operations, to take into account the work thus spent upon the atmosphere, and the restitution which has been made, since these precisely compensate for one another.

(2) The cylinder being removed, and placed on the impermeable stand K, *let the piston rise gradually, till, when it reaches a position* E_2F_2, *the temperature of the water and vapour is* T; *the same as that of the body* B.

[During this operation the fresh vapour continually formed requires heat to become latent; and, therefore, as the contents of the cylinder are protected from any accession of heat, their temperature sinks.]

(3) The cylinder being removed from K, and placed on B, *let the piston be pushed down, till, when it reaches the position* E_3F_3, *the quantity of heat evolved and abstracted by* B *amounts to that which, during the first operation, was taken from* A.

[Note of Nov. 5, 1881. The specification of this operation, with a view to the return to the primitive condition, intended as the conclusion to the four operations, is the only item in which Carnot's temporary and provisional assumption of the materiality of heat has effect. To exclude this hypothesis, Prof. James Thomson gave (see p. 161) the following corrected specification for the third operation ;—*Let the piston be pushed down, till it reaches a position* E_3F_3, *determined so as to fulfil the condition, that at the end of the fourth operation, the primitive temperature* S *shall be reached**:]

[During this operation the temperature of the contents of the cylinder is retained constantly at T^0, and all the latent heat of the vapour which is condensed into water at the same temperature, is given out to B.]

(4) The cylinder being removed from B, and placed on the impermeable stand, *let the piston be pushed down from* E_3F_3 *to its original position* EF.

[During this operation, the impermeable stand preventing any loss of heat, the temperature of the water and air must rise continually, till (since the quantity of heat evolved during the third operation was precisely equal to that which was

* [Note of Nov. 5, 1881. Maxwell has simplified the correction by beginning the cycle with Carnot's second operation, and completing it through his third, fourth, and first operations, with his third operation merely as follows :—

let the piston be pushed down to any position E_3F_3;

then Carnot's fourth operation altered to the following :—

let the piston be pushed down from E_3F_3 *until the temperature reaches its primitive value* S;

and lastly Carnot's first operation altered to the following :—

let the piston rise to its primitive position.]

previously absorbed), at the conclusion it reaches its primitive value, S, in virtue of Carnot's fundamental axiom.]

[Note of Nov. 5, 1881. With Prof. James Thomson's correction of operation (3), the words in virtue of "Carnot's Fundamental Axiom" must be replaced by "the condition fulfilled by operation (3)," in the description of the results of operation (4).]

16. At the conclusion of this cycle of operations* the total thermal agency has been the *letting down* of H units of heat from the body A, at the temperature S, to B, at the lower temperature T; and the aggregate of the mechanical effect has been a certain amount of *work produced*, since during the ascent of the piston in the first and second operations, the temperature of the water and vapour, and therefore the pressure of the vapour on the piston, was on the whole higher than during the descent, in the third and fourth operations. It remains for us actually to evaluate this aggregate amount of work performed; and for this purpose the following graphical method of representing the mechanical effect developed in the several operations, taken from Mons. Clapeyron's paper, is extremely convenient.

17. Let OX and OY be two lines at right angles to one another. Along OX measure off distances ON_1, N_1N_2, N_2N_3, N_3O, respectively proportional to the spaces described by the piston during the four successive operations described above; and, with reference to these four operations respectively, let the following constructions be made:—

(1) Along OY measure a length OA, to represent the pressure of the saturated vapour at the temperature S; and draw AA_1 parallel to OX, and let it meet an ordinate through N_1, in A_1.

(2) Draw a curve A_1PA such that, if ON represent, at any instant during the second operation, the distance of the piston from its primitive position, NP shall represent the pressure of the vapour at the same instant.

* In Carnot's work some perplexity is introduced with reference to the temperature of the water, which, in the operations he describes, is not brought back exactly to what it was at the commencement; but the difficulty which arises is explained by the author. No such difficulty occurs with reference to the cycle of operations described in the text, for which I am indebted to Mons. Clapeyron.

(3) Through A_2 draw A_2A_3 parallel to OX, and let it meet an ordinate through N_3 in A_3.

(4) Draw the curve A_3A such that the abscissa and ordinate of any point in it may represent respectively the distances of the piston from its primitive position, and the pressure of the

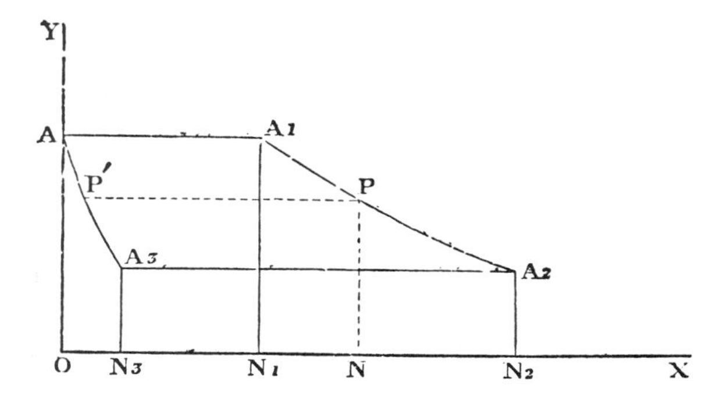

vapour, at each instant during the fourth operation. The last point of this curve must, according to Carnot's fundamental principle, coincide with A, since the piston is, at the end of the cycle of operations, again in its primitive position, and the pressure of the vapour is the same as it was at the beginning.

18. Let us now suppose that the lengths, ON_1, N_1N_2, N_2N_3, and N_3O, *represent numerically* the volumes of the spaces moved through by the piston during the successive operations. It follows that the mechanical effect obtained during the first operation will be *numerically represented* by the area AA_1N_1O; that is, the number of superficial units in this area will be equal to the number of "foot-pounds" of work performed by the ascending piston during the first operation. The work performed by the piston during the second operation will be similarly represented by the area $A_1A_2N_2N_1$. Again, during the third operation a certain amount of work is spent on the piston, which will be represented by the area $A_2A_3N_3N_2$; and lastly, during the fourth operation, work is spent in pushing the piston to an amount represented by the area A_3AON_3.

19. Hence the mechanical effect (represented by the area $OAA_1A_2N_2$) which was obtained during the first and second

operations, exceeds the work (represented by $N_2A_2A_3AO$) spent during the third and fourth, by an amount represented by the area of the quadrilateral figure $AA_1A_2A_3$; and, consequently, it only remains for us to evaluate this area, that we may determine the total mechanical effect gained in a complete cycle of operations. Now, from experimental data, at present nearly complete, as will be explained below, we may determine the length of the line AA_1 for the given temperature S, and a given absorption H, of heat, during the first operation; and the length of A_2A_3 for the given lower temperature T, and the evolution of the same quantity of heat during the fourth operation: and the curves A_1PA_2, $A_3P'A$ may be drawn as graphical representations of actual observations*. The figure being thus constructed, its area may be measured, and we are, therefore, in possession of a graphical method of determining the amount of mechanical effect to be obtained from any given thermal agency. As, however, it is merely the area of the figure which it is required to determine, it will not be necessary to be able to describe each of the curves A_1PA_2, $A_3P'A$, but it will be sufficient to know the difference of the abscissas corresponding to any equal ordinates in the two; and the following analytical method of completing the problem is the most convenient for leading to the actual numerical results.

20. Draw any line PP' parallel to OX, meeting the curvilineal sides of the quadrilateral in P and P'. Let ξ denote the length of this line, and p its distance from OX. The area of the figure, according to the integral calculus, will be denoted by the expression

$$\int_{p_3}^{p_1} \xi dp,$$

where p_1, and p_3 (the limits of integration indicated according to Fourier's notation) denote the lines OA, and N_3A_3, which represent respectively the pressures during the first and third operations. Now, by referring to the construction described above, we see that ξ is the difference of the volumes below the piston at corresponding instants of the second and fourth operations, or instants at which the saturated steam and the water in the cylinder have the same pressure p, and, consequently, the same temperature

* See Note at the end of this Paper.

which we may denote by t. Again, throughout the second operation the entire contents of the cylinder possess a greater amount of heat by H units than during the fourth; and, therefore, at any instant of the second operation there is as much more steam as contains H units of latent heat, than at the corresponding instant of the fourth operation. Hence, if k denote the latent heat in a unit of saturated steam at the temperature t, the volume of the steam at the two corresponding instants must differ by $\frac{H}{k}$. Now, if σ denote the ratio of the density of the steam to that of the water, the volume $\frac{H}{k}$ of steam will be formed from the volume $\sigma\frac{H}{k}$ of water; and, consequently, we have for the difference of volumes of the entire contents at the corresponding instants,

$$\xi = (1-\sigma)\frac{H}{k}.$$

Hence the expression for the area of the quadrilateral figure becomes

$$\int_{p_3}^{p_1} (1-\sigma)\frac{H}{k}\,dp.$$

Now, σ, k, and p, being quantities which depend upon the temperature, may be considered as functions of t; and it will be convenient to modify the integral so as to make t the independent variable. The limits will be from $t=T$ to $t=S$, and, if we denote by M the value of the integral, we have the expression

$$M = H\int_T^S (1-\sigma)\frac{dp}{kdt}\,dt \qquad (1),$$

for the total amount of mechanical effect gained by the operations described above.

21. If the interval of temperatures be extremely small; so small that $(1-\sigma)\,dp/kdt$ will not sensibly vary for values of t between T and S, the preceding expression becomes simply

$$M = (1-\sigma)\frac{dp}{kdt}\cdot H(S-T) \qquad (2).$$

This might, of course, have been obtained at once, by supposing the breadth of the quadrilateral figure AA_1A_2A to be extremely small compared with its length, and then taking for its area, as an approximate value, the product of the breadth into the line AA_1, or the line A_3A_2, or any line of intermediate magnitude.

The expression (2) is rigorously correct for any interval $S-T$, if the mean value of $(1-\sigma)\,dp/kdt$ for that interval be employed as the coefficient of $H\,(S-T)$.

CARNOT'S *Theory of the Air-Engine.*

22. In the ideal air-engine imagined by Carnot four operations performed upon a mass of air or gas enclosed in a closed vessel of variable volume, constitute a complete cycle, at the end of which the medium is left in its primitive physical condition; the construction being the same as that which was described above for the steam-engine, a body A, permanently retained at the temperature S, and B at the temperature T; an impermeable stand K; and a cylinder and piston, which, in this case, contains a mass of air at the temperature S, instead of water in the liquid state, at the beginning and end of a cycle of operations. The four successive operations are conducted in the following manner:—

(1) The cylinder is laid on the body A, so that the air in it is kept at the temperature S; and the piston is allowed to rise, performing work.

(2) The cylinder is placed on the impermeable stand K, so that its contents can neither gain nor lose heat, and the piston is allowed to rise farther, still performing work, till the temperature of the air sinks to T.

(3) The cylinder is placed on B, so that the air is retained at the temperature T, and the piston is pushed down till the air gives out to the body B as much heat as it had taken in from A, during the first operation.

[Note of Nov. 5, 1881. To eliminate the assumption of the materiality of heat, make Professor James Thomson's correction here also; as above in § 15; or take Maxwell's re-arrangement of the cycle described in the foot-note to § 15.]

(4) The cylinder is placed on K, so that no more heat can be taken in or given out, and the piston is pushed down to its primitive position.

23. *At the end of the fourth operation the temperature must have reached its primitive value S, in virtue of* CARNOT'S *axiom.*

24. Here, again, as in the former case, we observe that work is performed by the piston during the first two operations; and, during the third and fourth, work is spent upon it, but to a less amount, since the pressure is on the whole less during the third and fourth operations than during the first and second, on account of the temperature being lower. Thus, at the end of a complete cycle of operations, mechanical effect has been obtained; and the thermal agency from which it is drawn is the taking of a certain quantity of heat from A, and *letting it down*, through the medium of the engine, to the body B at a lower temperature.

25. To estimate the actual amount of effect thus obtained, it will be convenient to consider the alterations of volume of the mass of air in the several operations as extremely small. We may afterwards pass by the integral calculus, or, practically, by summation, to determine the mechanical effect whatever be the amplitudes of the different motions of the piston.

26. Let dq be the quantity of heat absorbed during the first operation, which is evolved again during the third; and let dv be the corresponding augmentation of volume which takes place while the temperature remains constant, as it does during the first operation*. The diminution of volume in the third

* Thus, $\frac{dq}{dv}$ will be the partial differential coefficient, with respect to v of that function of v and t, which expresses the quantity of heat that must be added to a mass of air when in a "standard" state (such as at the temperature zero, and under the atmospheric pressure), to bring it to the temperature t, and the volume v. That there is such a function, of two independent variables v and t, is merely an analytical expression of Carnot's fundamental axiom, as applied to a mass of air. The general principle may be analytically stated in the following terms:—If Mdv denote the accession of heat received by a mass of any kind, not possessing a destructible texture, when the volume is increased by dv, the temperature being

operation must be also equal to dv, or only differ from it by an infinitely small quantity of the second order. During the second operation we may suppose the volume to be increased by an infinitely small quantity ϕ; which will occasion a diminution of pressure, and a diminution of temperature, denoted respectively by ω and τ. During the fourth operation there will be a diminution of volume, and an increase of pressure and temperature, which can only differ, by infinitely small quantities of the second order, from the changes in the other direction, which took place in the second operation, and they also may, therefore, be denoted by ϕ, ω, and τ, respectively. The alteration of pressure, during the first and third operations, may at once be determined by means of Mariotte's law, since, in them, the temperature remains constant. Thus, if, at the commencement of the cycle, the volume and pressure be v and p, they will have become $v + dv$ and $pv/(v + dv)$ at the end of the first operation. Hence the diminution of pressure, during the first operation, is $p - pv/(v + dv)$ or $pdv/(v + dv)$; and, therefore, if we neglect infinitely small quantities of the second order, we have pdv/v for the diminution of pressure during the first operation; which, to the same degree of approximation, will be equal to the increase of pressure during the third. If $t + \tau$ and t be taken to denote the superior and inferior limits of temperature, we shall thus have for the volume, the temperature, and the pressure at the commencements of the four successive operations, and at the end of the cycle, the following values respectively:—

$$
\begin{array}{llll}
(1) & v, & t+\tau, & p; \\
(2) & v+dv, & t+\tau, & p\left(1-\dfrac{dv}{v}\right); \\
(3) & v+dv+\phi, & t, & p\left(1-\dfrac{dv}{v}\right)-\omega; \\
(4) & v+\phi, & t, & p-\omega; \\
(5) & v, & t+\tau, & p.
\end{array}
$$

kept constant, and if Ndt denote the amount of heat which must be supplied to raise the temperature by dt, without any alteration of volume; then $Mdv + Ndt$ must be the differential of a function of v and t. [Note of Nov. 5, 1881. In the corrected theory it is $(M - Jp)\,dv + Ndt$, that is a complete differential, not $Mdv + Ndt$. See *Dynamical Theory of Heat* (Art. XLVIII., below), § 20.]

Taking the mean of the pressures at the beginning and end of each operation, we find

$$(1)\quad p\left(1-\tfrac{1}{2}\frac{dv}{v}\right),\qquad (2)\quad p\left(1-\frac{dv}{v}\right)-\tfrac{1}{2}\omega,$$

$$(3)\quad p\left(1-\tfrac{1}{2}\frac{dv}{v}\right)-\omega,\qquad (4)\quad p-\tfrac{1}{2}\omega,$$

which, as we are neglecting infinitely small quantities of the second order, will be the expressions for the mean pressures during the four successive operations. Now, the mechanical effect gained or spent, during any of the operations, will be found by multiplying the mean pressure by the increase or diminution of volume which takes place; and we thus find

$$(1)\quad p\left(1-\tfrac{1}{2}\frac{dv}{v}\right)dv\qquad (2)\quad \left\{p\left(1-\frac{dv}{v}\right)-\tfrac{1}{2}\omega\right\}\phi$$

$$(3)\quad \left\{p\left(1-\tfrac{1}{2}\frac{dv}{v}\right)-\omega\right\}dv\qquad (4)\quad (p-\tfrac{1}{2}\omega)\,\phi$$

for the amounts gained during the first and second, and spent during the third and fourth operations; and hence, by addition and subtraction, we find

$$\omega dv - p\phi\frac{dv}{v},\quad \text{or}\quad (v\omega - p\phi)\frac{dv}{v},$$

for the aggregate amount of mechanical effect gained during the cycle of operations. It only remains for us to express this result in terms of dq and τ, on which the given thermal agency depends. For this purpose, we remark that ϕ and ω are alterations of volume and pressure which take place along with a change of temperature τ, and hence, by the laws of compressibility and expansion, we may establish a relation* between them in the following manner.

Let p_0 be the pressure of the mass of air when reduced to the temperature zero, and confined in a volume v_0; then, what-

* We might also nvestigate another relation, to express the fact that there is no accession or removal of heat during either the second or the fourth operation; but it will be seen that this will not affect the result in the text; although it would enable us to determine both ϕ and ω in terms of τ.

ever be v_0, the product p_0v_0 will, by the law of compressibility, remain constant; and, if the temperature be elevated from 0 to $t+\tau$, and the gas be allowed to expand freely without any change of pressure, its volume will be increased in the ratio of 1 to $1+E(t+\tau)$, where E is very nearly equal to ·00366 (the centigrade scale of the air-thermometer being referred to), whatever be the gas employed, according to the researches of Regnault and of Magnus on the expansion of gases by heat. If, now, the volume be altered arbitrarily with the temperature continually at $t+\tau$, the product of the pressure and volume will remain constant; and, therefore, we have

$$pv = p_0v_0\{1 + E(t+\tau)\}.$$

Similarly $$(p-\omega)(v+\phi) = p_0v_0\{1+Et\}.$$

Hence, by subtraction, we have

$$v\omega - p\phi + \omega\phi = p_0v_0E\tau,$$

or, neglecting the product $\omega\phi$,

$$v\omega - p\phi = p_0v_0E\tau.$$

Hence, the preceding expression for mechanical effect, gained in the cycle of operations, becomes $p_0v_0 \,.\, E\tau \,.\, dv/v$.

Or, as we may otherwise express it,

$$\frac{Ep_0v_0}{vdq/dv} \,.\, dq \,.\, \tau.$$

Hence, if we denote by M the mechanical effect due to H units of heat descending through the same interval τ, which might be obtained by repeating the cycle of operations described above, $\frac{H}{dq}$ times, we have

$$M = \frac{Ep_0v_0}{vdq/dv} \,.\, H\tau \qquad\qquad (3).$$

27. If the *amplitudes* of the operations had been finite, so as to give rise to an absorption of H units of heat during the first operation, and a lowering of temperature from S to T during the second, the amount of work obtained would have been found to be expressed by means of a double definite integral, thus* :—

* This result might have been obtained by applying the usual notation of the integral calculus to express the area of the curvilinear quadrilateral, which, ac-

$$M = \int_0^H dq \int_T^S dt \cdot \frac{Ep_0 v_0}{v dq/dv} \text{ or } M = Ep_0 v_0 \int_0^H \int_T^S \frac{1}{v} \frac{dv}{dq} \cdot dt dq; \ldots\ldots(4),$$

this second form being sometimes more convenient.

28. The preceding investigations, being founded on the approximate laws of compressibility and expansion (known as the law of Mariotte and Boyle, and the law of Dalton and Gay-Lussac), would require some slight modifications, to adapt them to cases in which the gaseous medium employed is such as to present sensible deviations from those laws. Regnault's very accurate experiments shew that the deviations are insensible, or very nearly so, for the ordinary gases at ordinary pressures; although they may be considerable for a medium, such as sulphurous acid, or carbonic acid under high pressure, which approaches the physical condition of a vapour at saturation; and therefore, in general, and especially in practical applications to real air-engines, it will be unnecessary to make any modification in the expressions. In cases where it may be necessary, there is no difficulty in making the modifications, when the requisite data are supplied by experiment.

29*. Either the steam-engine or the air-engine, according to the arrangements described above, gives all the mechanical effect that can possibly be obtained from the thermal agency employed. For it is clear, that, in either case, the operations may be performed in the reverse order, with every thermal and mechanical effect reversed. Thus, in the steam-engine, we may commence by placing the cylinder on the impermeable stand, allow the piston to rise, performing work, to the position E_3F_3; we may then place it on the body B, and allow it to rise, performing work, till it reaches E_2F_2; after that the cylinder may be placed again on

cording to Clapeyron's graphical construction, would be found to represent the entire mechanical effect gained in the cycle of operations of the air-engine. It is not necessary, however, to enter into the details of this investigation, as the formula (3), and the consequences derived from it, include the whole theory of the air-engine, in the best practical form; and the investigation of it which I have given in the text, will probably give as clear a view of the reasoning on which it is founded, as could be obtained by the graphical method, which, in this case, is not so valuable as it is from its simplicity in the case of the steam-engine.

* This paragraph is the demonstration referred to above, of the proposition stated in § 13; as it is readily seen that it is applicable to any conceivable kind of thermo-dynamic engine.

the impermeable stand, and the piston may be pushed down to E_1F_1; and, lastly, the cylinder being removed to the body A, the piston may be pushed down to its primitive position. In this inverse cycle of operations, a certain amount of work has been spent, precisely equal, as we readily see, to the amount of mechanical effect gained in the direct cycle described above; and heat has been abstracted from B, and deposited in the body A, at a higher temperature, to an amount precisely equal to that which, in the direct cycle, was *let down* from A to B. Hence it is impossible to have an engine which will derive more mechanical effect from the same thermal agency, than is obtained by the arrangement described above; since, if there could be such an engine, it might be employed to perform, as a part of its whole work, the inverse cycle of operations, upon an engine of the kind we have considered, and thus to continually restore the heat from B to A, which has descended from A to B for working itself; so that we should have a complex engine, giving a residual amount of mechanical effect without any thermal agency, or alteration of materials, which is an impossibility in nature. The same reasoning is applicable to the air-engine; and we conclude, generally, that any two engines, constructed on the principles laid down above, whether steam-engines with different liquids, an air-engine and a steam-engine, or two air-engines with different gases, must derive the same amount of mechanical effect from the same thermal agency.

30. Hence, by comparing the amounts of mechanical effect obtained by the steam-engine and the air-engine from the letting down of the H units of heat from A at the temperature $(t+\tau)$ to B at t, according to the expressions (2) and (3), we have

$$M = (1-\sigma)\frac{dp}{kdt} \cdot H\tau = \frac{Ep_0v_0}{vdq/dv} \cdot H\tau \ldots\ldots\ldots\ldots\ldots (5).$$

If we denote the coefficient of $H\tau$ in these equal expressions by μ, which may be called "Carnot's coefficient," we have

$$\mu = (1-\sigma)\frac{dp}{kdt} = \frac{Ep_0v_0}{vdq/dv} \ldots\ldots\ldots\ldots\ldots\ldots\ldots (6),$$

and we deduce the following very remarkable conclusions:—

(1) For the saturated vapours of all different liquids, at the same temperature, the value of $(1-\sigma)\frac{dp}{kdt}$ must be the same.

(2) For any different gaseous masses, at the same temperature, the value of $\frac{Ep_0 v_0}{vdq/dv}$ must be the same.

(3) The values of these expressions for saturated vapours and for gases, at the same temperature, must be the same.

31. No conclusion can be drawn *a priori* regarding the values of this coefficient μ for different temperatures, which can only be determined, or compared, by experiment. The results of a great variety of experiments, in different branches of physical science (Pneumatics and Acoustics), cited by Carnot and by Clapeyron, indicate that the values of μ for low temperatures exceed the values for higher temperatures; a result amply verified by the continuous series of experiments performed by Regnault on the saturated vapour of water for all temperatures from 0° to 230°, which, as we shall see below, give values for μ gradually diminishing from the inferior limit to the superior limit of temperature. When, by observation, μ has been determined as a function of the temperature, the amount of mechanical effect, M, deducible from H units of heat descending from a body at the temperature S to a body at the temperature T, may be calculated from the expression,

$$M = H\int_T^S \mu dt \text{.........................} (7),$$

which is, in fact, what either of the equations (1) for the steam-engine, or (4) for the air-engine, becomes, when the notation μ, for Carnot's multiplier, is introduced.

The values of this integral may be practically obtained, in the most convenient manner, by first determining, from observation, the mean values of μ for the successive degrees of the thermometric scale, and then adding the values for all the degrees within the limits of the extreme temperatures S and T*.

* The results of these investigations are exhibited in Tables I. and II. below.

32. The complete theoretical investigation of the motive power of heat is thus reduced to the experimental determination of the coefficient μ; and may be considered as perfect, when, by any series of experimental researches whatever, we can find a value of μ for every temperature within practical limits. The special character of the experimental researches, whether with reference to gases, or with reference to vapours, necessary and sufficient for this object, is defined and restricted in the most precise manner, by the expressions (6) for μ, given above.

33. The object of Regnault's great work, referred to in the title of this paper, is the experimental determination of the various physical elements of the steam-engine; and when it is complete, it will furnish all the *data* necessary for the calculation of μ. The valuable researches already published in a first part of that work, make known the latent heat of a given weight, and the pressure, of saturated steam for all temperatures between 0° and 230° cent. of the air-thermometer. Besides these data, however, the density of saturated vapour must be known, in order that k, the latent heat of a unit of volume, may be calculated from Regnault's determination of the latent heat of a given weight*. Between the limits of 0° and 100°, it is probable, from various experiments which have been made, that the density of vapour follows very closely the simple laws which are so accurately verified by the ordinary gases†; and thus it may be calculated from Regnault's table giving the pressure at any temperature within those limits. Nothing as yet is known with accuracy as to the density of saturated steam between 100° and 230°, and we must be contented at present to estimate it by calculation from Regnault's table of pressures; although, when accurate experimental researches on the subject shall have been

* It is, comparatively speaking, of little consequence to know accurately the value of σ, for the factor $(1-\sigma)$ of the expression for μ, since it is so small (being less than $\frac{1}{1700}$ for all temperatures between 0° and 100°) that, unless all the data are known with more accuracy than we can count upon at present, we might neglect it altogether, and take dp/kdt simply, as the expression for μ, without committing any error of important magnitude.

† This is well established, within the ordinary atmospheric limits, in Regnault's Études Météorologiques, in the *Annales de Chimie*, Vol. xv., 1846.

made, considerable deviations from the laws of Boyle and Dalton, on which this calculation is founded, may be discovered.

34. Such are the experimental data on which the mean values of μ for the successive degrees of the air-thermometer, from 0° to 230°, at present laid before the Royal Society, is founded. The unit of length adopted is the English foot; the unit of weight, the pound; the unit of work, a "foot-pound;" and the unit of heat that quantity which, when added to a pound of water at 0°, will produce an elevation of 1° in temperature. The mean value of μ for any degree is found to a sufficient degree of approximation, by taking, in place of σ, dp/dt, and k; in the expression

$$(1-\sigma)\frac{dp}{kdt};$$

the mean values of those elements; or, what is equivalent to the corresponding accuracy of approximation, by taking, in place of σ and k respectively, the mean of the values of those elements for the limits of temperature, and in place of dp/dt, the difference of the values of p, at the same limits.

35. In Regnault's work (at the end of the eighth Mémoire), a table of the pressures of saturated steam for the successive temperatures 0°, 1°, 2°,... 230°, expressed in millimetres of mercury, is given. On account of the units adopted in this paper, these pressures must be estimated in pounds on the square foot, which we may do by multiplying each number of millimetres by 2·7896, the weight in pounds of a sheet of mercury, one millimetre thick, and a square foot in area.

36. The value of k, the latent heat of a cubic foot, for any temperature t, is found from λ, the latent heat of a pound of saturated steam, by the equation

$$k=\frac{p}{760}\cdot\frac{1+\cdot 00366\times 100}{1+\cdot 00366\times t}.\times\cdot 036869^{*}.\lambda,$$

* It appears that the vol. of 1 kilog. must be 1·69076 according to the data here assumed.

The density of saturated steam at 100° is taken as $\frac{1}{1693\cdot 5}$ of that of water at its maximum. Rankins takes it as $\frac{1}{1696}$.

where p denotes the pressure in millimetres, and λ the latent heat of a pound of saturated steam; the values of λ being calculated by the empirical formula*

$$\lambda = (606{\cdot}5 + 0{\cdot}305t) - (t + {\cdot}00002t^2 + 0{\cdot}0000003t^3),$$

given by Regnault as representing, between the extreme limits of his observations, the latent heat of a unit weight of saturated steam.

Explanation of Table I.

37. The mean values of μ for the first, for the eleventh, for the twenty-first, and so on, up to the 231st† degree of the air-thermometer, have been calculated in the manner explained in the preceding paragraphs. These, and interpolated results, which must agree with what would have been obtained, by direct calculation from Regnault's data, to three significant places of figures (and even for the temperatures between 0^0 and 100^0, the experimental data do not justify us in relying on any of the results to a greater degree of accuracy), are exhibited in Table I.

To find the amount of mechanical effect due to a unit of heat, descending from a body at a temperature S to a body at T, if these numbers be integers, we have merely to add the values of μ in Table I. corresponding to the successive numbers.

$$T+1,\ T+2, \ldots\ldots S-2,\ S-1.$$

Explanation of Table II.

38. The calculation of the mechanical effect, in any case, which might always be effected in the manner described in § 37

* The part of this expression in the first vinculum (see Regnault, end of ninth Mémoire) is what is known as "the total heat" of a pound of steam, or the amount of heat necessary to convert a pound of water at 0^0 into a pound of saturated steam at t^0; which, according to "Watt's law," thus approximately verified, would be constant. The second part, which would consist of the single term t, if the specific heat of water were constant for all temperatures, is the number of thermic units necessary to raise the temperature of a pound of water from 0^0 to t^0, and expresses empirically the results of Regnault's experiments on the specific heat of water (see end of the tenth Mémoire), described in the work already referred to.

† In strictness, the 230th is the last degree for which the experimental data are complete; but the data for the 231st may readily be assumed in a sufficiently satisfactory manner.

(with the proper modification for fractions of degrees, when necessary), is much simplified by the use of Table II., where the first number of Table I., the sum of the first and second, the sum of the first three, the sum of the first four, and so on, are successively exhibited. The sums thus tabulated are the values of the integrals

$$\int_0^1 \mu dt, \quad \int_0^2 \mu dt, \quad \int_0^3 \mu dt, \ldots\ldots \int_0^{231} \mu dt;$$

and, if we denote $\int_0^t \mu dt$ by the letter M, Table II. may be regarded as a table of the value of M.

To find the amount of mechanical effect due to a unit of heat descending from a body at a temperature S to a body at T, if these numbers be integers, we have merely to subtract the value of M, for the number T, from the value for the number S, given in Table II.

TABLE I.* *Mean Values of μ for the successive Degrees of the Air-Thermometer from 0° to 230°.*

	μ		μ		μ		μ		μ
1°	4·960	48°	4·366	94°	3·889	140°	3·549	186°	3·309
2	4·946	49	4·355	95	3·880	141	3·543	187	3·304
3	4·932	50	4·343	96	3·871	142	3·537	188	3·300
4	4·918	51	4·331	97	3·863	143	3·531	189	3·295
5	4·905	52	4·319	98	3·854	144	3·525	190	3·291
6	4·892	53	4·308	99	3·845	145	3·519	191	3·287
7	4·878	54	4·296	100	3·837	146	3·513	192	3·282
8	4·865	55	4·285	101	3·829	147	3·507	193	3·278
9	4·852	56	4·273	102	3·820	148	3·501	194	3·274
10	4·839	57	4·262	103	3·812	149	3·495	195	3·269
11	4·826	58	4·250	104	3·804	150	3·490	196	3·265
12	4·812	59	4·239	105	3·796	151	3·484	197	3·261
13	4·799	60	4·227	106	3·788	152	3·479	198	3·257
14	4·786	61	4·216	107	3·780	153	3·473	199	3·253
15	4·773	62	4·205	108	3·772	154	3·468	200	3·249
16	4·760	63	4·194	109	3·764	155	3·462	201	3·245
17	4·747	64	4·183	110	3·757	156	3·457	202	3·241
18	4·735	65	4·172	111	3·749	157	3·451	203	3·237
19	4·722	66	4·161	112	3·741	158	3·446	204	3·233
20	4·709	67	4·150	113	3·734	159	3·440	205	3·229
21	4·697	68	4·140	114	3·726	160	3·435	206	3·225
22	4·684	69	4·129	115	3·719	161	3·430	207	3·221
23	4·672	70	4·119	116	3·712	162	3·424	208	3·217
24	4·659	71	4·109	117	3·704	163	3·419	209	3·213
25	4·646	72	4·098	118	3·697	164	3·414	210	3·210
26	4·634	73	4·088	119	3·689	165	3·409	211	3·206
27	4·621	74	4·078	120	3·682	166	3·404	212	3·202
28	4·609	75	4·067	121	3·675	167	3·399	213	3·198
29	4·596	76	4·057	122	3·668	168	3·394	214	3·195
30	4·584	77	4·047	123	3·661	169	3·389	215	3·191
31	4·572	78	4·037	124	3·654	170	3·384	216	3·188
32	4·559	79	4·028	125	3·647	171	3·380	217	3·184
33	4·547	80	4·018	126	3·640	172	3·375	218	3·180
34	4·535	81	4·009	127	3·633	173	3·370	219	3·177
35	4·522	82	3·999	128	3·627	174	3·365	220	3·173
36	4·510	83	3·990	129	3·620	175	3·361	221	3·169
37	4·498	84	3·980	130	3·614	176	3·356	222	3·165
38	4·486	85	3·971	131	3·607	177	3·351	223	3·162
39	4·474	86	3·961	132	3·601	178	3·346	224	3·158
40	4·462	87	3·952	133	3·594	179	3·342	225	3·155
41	4·450	88	3·943	134	3·586	180	3·337	226	3·151
42	4·438	89	3·934	135	3·579	181	3·332	227	3·148
43	4·426	90	3·925	136	3·573	182	3·328	228	3·144
44	4·414	91	3·916	137	3·567	183	3·323	229	3·141
45	4·402	92	3·907	138	3·561	184	3·318	230	3·137
46	4·390	93	3·898	139	3·555	185	3·314	231	3·134
47	4·378								

* The numbers here tabulated may also be regarded as *the actual values of μ for* $t=\frac{1}{2}$, $t=1\frac{1}{2}$, $t=2\frac{1}{2}$, $t=3\frac{1}{2}$, &c.

TABLE II. *Mechanical Effect in Foot-Pounds due to a Thermic Unit Centigrade, passing from a body, at any Temperature less than 230° to a body at 0°.*

Superior Limit of Tempe-rature.	Mechanical Effect.	Superior Limit of Tempe-rature.	Mechanical Effect.	Superior Limit of Tempe-rature.	Mechanical Effect.	Superior Limit of Tempe-rature.	Mechanical Effect.	Superior Limit of Tempe-rature.	Mechanical Effect.
	Foot-Pounds.		Foot-Pounds.		Foot-Pounds.		Foot-Pounds.		Foot-Pounds.
1°	4·960	48°	223·487	94°	412·545	140°	582·981	186°	740·310
2	9·906	49	227·842	95	416·425	141	586·524	187	743·614
3	14·838	50	232·185	96	420·296	142	590·061	188	746·914
4	19·756	51	236·516	97	424·159	143	593·592	189	750·209
5	24·661	52	240·835	98	428·013	144	597·117	190	753·500
6	29·553	53	245·143	99	431·858	145	600·636	191	756·787
7	34·431	54	249·439	100	435·695	146	604·099	192	760·069
8	39·296	55	253·724	101	439·524	147	607·656	193	763·347
9	44·148	56	257·997	102	443·344	148	611·157	194	766·621
10	48·987	57	262·259	103	447·156	149	614·652	195	769·890
11	53·813	58	266·509	104	450·960	150	618·142	196	773·155
12	58·625	59	270·748	105	454·756	151	621·626	197	776·416
13	63·424	60	274·975	106	458·544	152	625·105	198	779·673
14	68·210	61	279·191	107	462·324	153	628·578	199	782·926
15	72·983	62	283·396	108	466·096	154	632·046	200	786·175
16	77·743	63	287·590	109	469·860	155	635·508	201	789·420
17	82·490	64	291·773	110	473·617	156	638·965	202	792·661
18	87·225	65	295·945	111	477·366	157	642·416	203	795·898
19	91·947	66	300·106	112	481·107	158	645·862	204	799·131
20	96·656	67	304·256	113	484·841	159	649·302	205	802·360
21	101·353	68	308·396	114	488·567	160	652·737	206	805·585
22	106·037	69	312·525	115	492·286	161	656·167	207	808·806
23	110·709	70	316·644	116	495·998	162	659·591	208	812·023
24	115·368	71	320·752	117	499·702	163	663·010	209	815·236
25	120·014	72	324·851	118	503·399	164	666·424	210	818·446
26	124·648	73	328·939	119	507·088	165	669·833	211	821·652
27	129·269	74	333·017	120	510·770	166	673·237	212	824·854
28	133·878	75	337·084	121	514·445	167	676·636	213	828·052
29	138·474	76	341·141	122	518·113	168	680·030	214	831·247
30	143·058	77	345·188	123	521·174	169	683·419	215	834·438
31	147·630	78	349·225	124	525·428	170	686·803	216	837·626
32	152·189	79	353·253	125	529·075	171	690·183	217	840·810
33	156·736	80	357·271	126	532·715	172	693·558	218	843·990
34	161·271	81	361·280	127	536·348	173	696·928	219	847·167
35	165·793	82	365·279	128	539·975	174	700·293	220	850·340
36	170·303	83	369·269	129	543·595	175	703·654	221	853·509
37	174·801	84	373·249	130	547·209	176	707·010	222	856·674
38	179·287	85	377·220	131	550·816	177	710·361	223	859·836
39	183·761	86	381·181	132	554·417	178	713·707	224	862·994
40	188·223	87	385·133	133	558·051	179	717·049	225	866·149
41	192·673	88	389·076	134	561·597	180	720·386	226	869·300
42	197·111	89	393·010	135	565·176	181	723·718	227	872·448
43	201·537	90	396·935	136	568·749	182	727·046	228	875·592
44	205·951	91	400·851	137	572·316	183	730·369	229	878·733
45	210·353	92	404·758	138	575·877	184	733·687	230	881·870
46	214·743	93	408·656	139	579·432	185	737·001	231	885·004
47	219·121								

[*Editor's Note:* Material has been omitted at this point.]

5

Reprinted from pp. 1–12 of *Philos. Mag.*, ser. 4, 2, 1–20, 102–119 (1851)

ON THE MOVING FORCE OF HEAT, AND THE LAWS REGARDING THE NATURE OF HEAT ITSELF WHICH ARE DEDUCIBLE THEREFROM

R. Clausius*

THE steam-engine having furnished us with a means of converting heat into a motive power, and our thoughts being thereby led to regard a certain quantity of work as an equivalent for the amount of heat expended in its production, the idea of establishing theoretically some fixed relation between a quantity of heat and the quantity of work which it can possibly produce, from which relation conclusions regarding the nature of heat itself might be deduced, naturally presents itself. Already, indeed, have many instructive experiments been made with this view; I believe, however, that they have not exhausted the subject, but that, on the contrary, it merits the continued attention of physicists; partly because weighty objections lie in the way of the conclusions already drawn, and partly because other conclusions, which might render efficient aid towards establishing and completing the theory of heat, remain either entirely unnoticed, or have not as yet found sufficiently distinct expression.

The most important investigation in connexion with this subject is that of S. Carnot†. Later still, the ideas of this author have been represented analytically in a very able manner by Clapeyron‡. Carnot proves that whenever work is produced by heat, and a permanent alteration of the body in action does not at the same time take place, a certain quantity of heat passes

* Translated from Poggendorff's *Annalen*, vol. lxxix. p. 368.

† *Reflexions sur la puissance motrice du feu, et sur les Machines propres à développer ceite puissance*, par S. Carnot. Paris, 1824.

‡ *Journ. de l'École Polytechnique*, vol. xix. (1834); and Taylor's Scientific Memoirs, Part III. p. 347.

from a warm body to a cold one; for example, the vapour which is generated in the boiler of a steam-engine, and passes thence to the condenser where it is precipitated, carries heat from the fireplace to the condenser. This *transmission* Carnot regards as the change of heat corresponding to the work produced. He says expressly, that *no heat is lost* in the process, that the quantity remains unchanged; and he adds, "This is a fact which has never been disputed; it is first assumed without investigation, and then confirmed by various calorimetric experiments. To deny it, would be to reject the entire theory of heat, of which it forms the principal foundation."

I am not, however, sure that the assertion, that in the production of work a loss of heat never occurs, is sufficiently established by experiment. Perhaps the contrary might be asserted with greater justice; that although no such loss may have been directly proved, still other facts render it exceedingly probable that a loss occurs. If we assume that heat, like matter, cannot be lessened in quantity, we must also assume that it cannot be increased; but it is almost impossible to explain the ascension of temperature brought about by friction otherwise than by assuming an actual increase of heat. The careful experiments of Joule, who developed heat in various ways by the application of mechanical force, establish almost to a certainty, not only the possibility of increasing the quantity of heat, but also the fact that the newly-produced heat is proportional to the work expended in its production. It may be remarked further, that many facts have lately transpired which tend to overthrow the hypothesis that heat is itself a body, and to prove that it consists in a motion of the ultimate particles of bodies. If this be so, the general principles of mechanics may be applied to heat; this motion may be converted into work, the loss of *vis viva* in each particular case being proportional to the quantity of work produced.

These circumstances, of which Carnot was also well aware, and the importance of which he expressly admitted, pressingly demand a comparison between heat and work, to be undertaken with reference to the divergent assumption that the production of work is not only due to an alteration in the *distribution* of heat, but to an actual *consumption* thereof; and inversely, that by the consumption of work heat may be *produced*.

In a recent memoir by Holtzmann*, it seemed at first as if the author intended to regard the subject from this latter point of view. He says (p. 7), "the effect of the heat which has been communicated to the gas is either an increase of temperature

* *Ueber die Wärme und Elasticität der Gase und Dämpfe*, von C. Holtzmann. Manheim, 1845. Also Taylor's Scientific Memoirs, Part XIV. p. 189.

combined with an increase of elasticity, or a mechanical work, or a combination of both ; a mechanical work being the equivalent for an increase of temperature. Heat can only be measured by its effects; and of the two effects mentioned, mechanical work is peculiarly applicable here, and shall therefore be chosen as a standard in the following investigation. I name a unit of heat, the quantity which, on being communicated to any gas, is able to produce the quantity of work *a* ; or to speak more definitely, which is able to raise *a* kilogrammes to a height of one metre." Afterwards, at page 12, he determines the numerical value of the constant *a*, according to the method of Meyer*, and obtains a number which completely agrees with that obtained in a manner totally different by Joule. In carrying out the theory, however, that is, in developing the equations by means of which his conclusions are arrived at, he proceeds in a manner similar to Clapeyron, so that the assumption that the quantity of heat is constant is still tacitly retained.

The difference between both ways of regarding the subject has been laid hold of with much greater clearness by W. Thomson, who has applied the recent discoveries of Regnault on the tension and latent heat of steam to the completing of the memoir of Carnot†. Thomson mentions distinctly the obstacles which lie in the way of an unconditional acceptance of Carnot's theory, referring particularly to the investigations of Joule, and dwelling on one principal objection to which the theory is liable. If it be even granted that the production of work, where the body in action remains in the same state after the production as before, is in all cases accompanied by a transmission of heat from a warm body to a cold one, it does not follow that by every such transmission work is produced, for the heat may be carried over by simple conduction; and in all such cases, if the transmission alone were the true equivalent of the work performed, an absolute loss of mechanical force must take place in nature, which is hardly conceivable. Notwithstanding this, however, he arrives at the conclusion, that in the present state of science the principle assumed by Carnot is the most probable foundation for an investigation on the moving force of heat. He says, "If we forsake this principle, we stumble immediately on innumerable other difficulties, which, without further experimental investigations, and an entirely new erection of the theory of heat, are altogether insurmountable."

I believe, nevertheless, that we ought not to suffer ourselves to be daunted by these difficulties; but that, on the contrary, we must look steadfastly into this theory which calls heat a motion, as in this way alone can we arrive at the means of establishing

* *Ann. der Chim und Pharm.*, vol. xlii. p. 239.

† Transactions of the Royal Society of Edinburgh, vol. xvi.

it or refuting it. Besides this, I do not imagine that the difficulties are so great as Thomson considers them to be; for although a certain alteration in our way of regarding the subject is necessary, still I find that this is in no case contradicted by *proved facts.* It is not even requisite to cast the theory of Carnot overboard; a thing difficult to be resolved upon, inasmuch as experience to a certain extent has shown a surprising coincidence therewith. On a nearer view of the case, we find that the new theory is opposed, not to the real fundamental principle of Carnot, but to the addition "no heat is lost;" for it is quite possible that in the production of work both may take place at the same time; a certain portion of heat may be consumed, and a further portion transmitted from a warm body to a cold one; and both portions may stand in a certain definite relation to the quantity of work produced. This will be made plainer as we proceed; and it will be moreover shown, that the inferences to be drawn from both assumptions may not only exist together, but that they mutually support each other.

1. *Deductions from the principle of the equivalence of heat and work.*

We shall forbear entering at present on the nature of the motion which may be supposed to exist within a body, and shall assume generally that a motion of the particles does exist, and that heat is the measure of their *vis viva.* Or yet more general, we shall merely lay down one maxim which is founded on the above assumption:—

In all cases where work is produced by heat, a quantity of heat proportional to the work done is expended; and inversely, by the expenditure of a like quantity of work, the same amount of heat may be produced.

Before passing on to the mathematical treatment of this maxim, a few of its more immediate consequences may be noticed, which have an influence on our entire notions as to heat, and which are capable of being understood, without entering upon the more definite proofs by calculation which are introduced further on.

We often hear of the *total heat* of bodies, and of gases and vapours in particular, this term being meant to express the sum of the sensible and latent heat. It is assumed that this depends solely upon the present condition of the body under consideration; so that when all other physical properties thereof, its temperature, density, &c. are known, the total quantity of heat which the body contains may also be accurately determined. According to the above maxim, however, this assumption cannot be admitted. If a body in a certain state, for instance a quantity of gas at the temperature t_0 and volume v_0, be subjected to various alterations as regards temperature and volume, and brought at the conclusion into its original state, the sum of its sensible and latent heats must, according to the above assumption, be the same as before; hence, if during any portion of the process heat be communicated from without, the quantity thus received must be given off again during some other portion of the process. With every alteration of volume, however, a certain quantity of work is either produced or expended by the gas; for by its expansion an outward pressure is forced back, and on the other hand, compression can only be effected by the advance of an outward pressure. If, therefore, alteration of volume be among the changes which the gas has undergone, work must be produced and expended. It is not, however, necessary that at the conclusion, when the original condition of the gas is again established, the entire amount of work produced should be exactly equal to the amount expended, the one thus balancing the other; an excess of one or the other will be present if the compression has taken place at a lower or a higher temperature than the expansion, as shall be proved more strictly further on. This excess of produced or expended work must, according to the maxim, correspond to a proportionate excess of expended or produced heat, and hence the amount of heat refunded by the gas cannot be the same as that which it has received.

There is still another way of exhibiting this divergence of our maxim from the common assumption as to the *total heat* of bodies. When a gas at t_0 and v_0 is to be brought to the higher temperature t_1 and the greater volume v_1, the quantity of heat necessary to effect this would, according to the usual hypothesis, be quite independent of the manner in which it is communicated. By the above maxim, however, this quantity would be different according as the gas is first heated at the constant volume v_0 and then permitted to expand at the constant temperature t_1, or first expanded at the temperature t_0 and afterwards heated to t_1; the quantity of heat varying in all cases with the manner in which the alterations succeed each other.

In like manner, when a quantity of water at the temperature t_0 is to be converted into vapour of the temperature t_1 and the volume v_1, it will make a difference in the amount of heat necessary if the water be heated first to t_1 and then suffered to evaporate, or if it be suffered to evaporate by t_0 and the vapour heated afterwards to t_1; or finally, if the evaporation take place at any intermediate temperature.

From this and from the immediate consideration of the maxim, we can form a notion as to the light in which *latent* heat must be regarded. Referring again to the last example, we distinguish in the quantity of heat imparted to the water during the change the *sensible* heat and the *latent* heat. Only the former of these, however, must we regard as present in the produced steam; the second is, not only as its name imports, hidden from our perceptions, but has actually *no existence*; during the alteration it has been *converted into work*.

We must introduce another distinction still as regards the heat expended. The work produced is of a twofold nature. In the first place, a certain quantity of work is necessary to overcome the mutual attraction of the particles, and to separate them to the distance which they occupy in a state of vapour. Secondly, the vapour during its development must, in order to procure room for itself, force back an outer pressure. We shall name the former of these *interior work*, and the latter *exterior work*, and shall distribute the latent heat also under the same two heads.

With regard to the *interior* work, it can make no difference whether the evaporation takes place at t_0 or at t_1, or at any other intermediate temperature, inasmuch as the attraction of the particles must be regarded as invariable*. The *exterior* work, on the contrary, is regulated by the pressure, and therefore by the temperature also. These remarks are not restricted to the example we have given, but are of general application; and when it was stated above, that the quantity of heat necessary to bring a body from one condition into another depended, not upon the state of the body at the beginning and the end alone, but upon the manner in which the alterations had been carried on throughout, this statement had reference to that portion only of the *latent heat* which corresponds to the *exterior* work. The remainder of the latent heat and the entire amount of sensible heat are independent of the manner in which the alteration is effected.

When the vapour of water at t_1 and v_1 is reconverted into water at t_0, the reverse occurs. Work is here *expended*, inasmuch as the particles again yield to their attraction, and the outer pressure once more advances. In this case, therefore, heat must be produced; and the *sensible heat* which here exhibits itself does not come from any retreat in which it was previously concealed, but is *newly produced*. It is not necessary that the heat developed by this reverse process should be equal to that consumed by the other; that portion which corresponds to the *exterior* work may be greater or less according to circumstances.

We shall now turn to the mathematical treatment of the subject, confining ourselves, however, to the consideration of per-

* It must not be objected here that the cohesion of the water at t_1 is less than at t_0, and hence requires a less amount of work to overcome it. The lessening of the cohesion implies a certain work performed by the warming of the water as water, and this must be added to that produced by evaporation. From this it follows, that of the heat which the water receives from without, only one portion must be regarded as sensible, while the other portion goes to loosen the cohesion. This view is in harmony with the fact, that water possesses a so much greater specific heat than ice, and probably than steam also.

manent gases, and of vapours at their maximum density; as besides possessing the greatest interest, our superior knowledge of these recommends them as best suited to the calculus. It will, however, be easy to see how the maxim may be applied to other cases also.

Let a certain quantity of *permanent gas,* say a unit of weight, be given. To determine its present condition, three quantities are necessary; the pressure under which it exists, its volume, and its temperature. These quantities stand to each other in a relation of mutual dependence, which, by a union of the laws of Mariotte and Gay-Lussac*, is expressed in the following equation:

$$pv = R(a+t), \quad . \quad . \quad . \quad . \quad . \quad . \quad (I.)$$

where p, v, *and* t express the pressure, volume, and temperature of the gas in its present state, a a constant equal for all gases, and R also a constant, which is fully expressed thus, $\frac{p_0 v_0}{a+t_0}$, where p_0, v_0, and t_0 express contemporaneous values of the above three quantities for any other condition of the gas. This last constant is therefore different for different gases, being inversely proportional to the specific weight of each.

It must be remarked, that Regnault has recently proved, by a series of very careful experiments, that this law is not in all strictness correct. The deviations, however, for the permanent gases are very small, and exhibit themselves principally in those cases where the gas admits of condensation. From this it would seem to follow, that the more distant, as regards pressure and temperature, a gas is from its point of condensation, the more correct will be the law. Its accuracy for permanent gases in their common state is so great, that it may be regarded as perfect; for every gas a limit may be imagined, up to which the law is also perfectly true; and in the following pages, where the permanent gases are treated as such, we shall assume the existence of this ideal condition.

The value $\frac{1}{a}$ for atmospheric air is found by the experiments both of Magnus and Regnault to be $=0{\cdot}003665$, the temperature being expressed by the centesimal scale reckoned from the freezing-point upwards. The gases, however, as already mentioned, not following strictly the law of M. and G., we do not always obtain the same value for $\frac{1}{a}$ when the experiment is repeated under different circumstances. The number given above is true for the case when the air is taken at a temperature of 0° under the pressure of *one* atmosphere, heated to a temperature

* This shall be expressed in future briefly thus—the law of M. and G.; and the law of Mariotte alone thus—the law of M.

of 100°, and the increase of expansive force observed. If, however, the pressure be allowed to remain constant, and the increase of volume observed, we obtain the somewhat higher value 0·003670. Further, the values increase when the experiments are made under a pressure exceeding that of the atmosphere, and decrease when the pressure is less. It is clear from this, that the exact value for the ideal condition, where the differences pointed out would of course disappear, cannot be ascertained. It is certain, however, that the number 0·003665 is not far from the truth, especially as it very nearly agrees with the value found for hydrogen, which, perhaps of all gases, approaches nearest the ideal condition. Retaining, therefore, the above value for $\frac{1}{a}$, we have

$$a=273.$$

One of the quantitie equation (I.), for instance p, may be regarded as a function of the two others; the latter will then be the independent variables which determine the condition of the gas. We will now endeavour to ascertain in what manner the quantities which relate to the *amount of heat* depend upon v and t.

When any body whatever changes its volume, the change is always accompanied by a mechanical work produced or expended. In most cases, however, it is impossible to determine this with accuracy, because an unknown *interior* work usually goes on at the same time with the *exterior*. To avoid this difficulty, Carnot adopted the ingenious contrivance before alluded to: he allowed the body to undergo various changes, and finally brought it into its primitive state; hence if by any of the changes *interior* work was produced, this was sure to be exactly nullified by some other change; and it was certain that the quantity of *exterior* work which remained over and above was the total quantity produced. Clapeyron has made this very evident by means of a diagram: we propose following his method with permanent gases in the first instance, introducing, however, some slight modifications rendered necessary by our maxim.

In the annexed figure let oe represent the volume, and ea the pressure of the unit weight of gas when its temperature is t; let us suppose the gas to be contained in an expansible bag, with which, however, no exchange of heat is possible. If the gas be permitted to expand, no new heat being added, the temperature will fall. To avoid

Fig. 1.

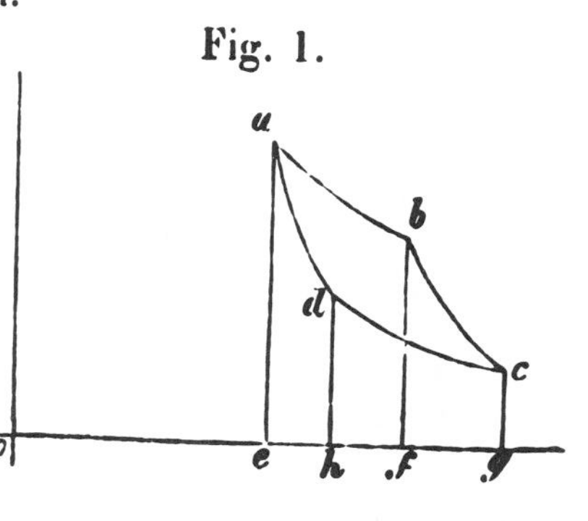

this, let the bag during the expansion be brought into contact with a body A of the temperature t, from which it shall receive heat sufficient to preserve it constant at the same temperature. While this expansion by constant temperature proceeds, the pressure decreases according to the law of M., and may be represented by the ordinate of a curve *ab*, which is a portion of an equilateral hyperbola. When the gas has increased in volume from *oe* to *of*, let the body A be taken away, and the expansion allowed to proceed still further without the addition of heat; the temperature will now sink, and the pressure consequently grow less as before. Let the law according to which this proceeds be represented by the curve *bc*. When the volume of the gas has increased from *of* to *og*, and its temperature is lowered from t to τ, let a pressure be commenced to bring it back to its original condition. Were the gas left to itself, its temperature would now rise; this, however, must be avoided by bringing it into contact with the body B at the temperature τ, to which any excess of heat will be immediately imparted, the gas being thus preserved constantly at τ. Let the compression continue till the volume has receded to h, it being so arranged that the decrease of volume indicated by the remaining portion *he* shall be just sufficient to raise the gas from τ to t, if during this decrease it gives out no heat. By the first compression the pressure increases according to the law of M., and may be represented by a portion *cd* of another equilateral hyperbola. At the end the increase is quicker, and may be represented by the curve *da*. This curve must terminate exactly in *a*; for as the volume and temperature at the end of the operation have again attained their original values, this must also be the case with the pressure, which is a function of both. The gas will therefore be found in precisely the same condition as at the commencement.

In seeking to determine the amount of work performed by these alterations, it will be necessary, for the reasons before assigned, to direct our attention to the *exterior* work alone. During the expansion, the gas *produces* a work expressed by the integral of the product of the differential of the volume into the corresponding pressure, which product is represented geometrically by the quadrilaterals *ea*, *bf* and *fbcg*. During the compression, however, work will be *expended*, which is represented by the quadrilaterals *gcdh* and *hdae*. The excess of the former work above the latter is to be regarded as the entire work produced by the alterations, and this is represented by the quadrilateral *abcd*.

If the foregoing process be reversed, we obtain at the conclusion the same quantity *abcd* as the excess of the work *expended* over that *produced*.

In applying the foregoing considerations analytically, we will assume that the various alterations which the gas has undergone have been *infinitely small.* We can then consider the curves before mentioned to be straight lines, as shown in the accompanying figure. In determining its superficial content, the quadrilateral *abcd* may be regarded as a parallelogram, for the error in this case can only amount to a differential of the *third* order, while the area itself is a differential of the *second* order. The latter may therefore be expressed by the product *ef.bk,* where *k* marks the point at which the ordinate *bf* cuts the lower side of the parallelogram. The quantity *bk* is the increase of pressure due to the raising of the constant volume *of* from τ to t, that is to say, due to the differential $t-\tau=dt$. This quantity can be expressed in terms of v and t by means of equation (I.), as follows:

Fig. 2.

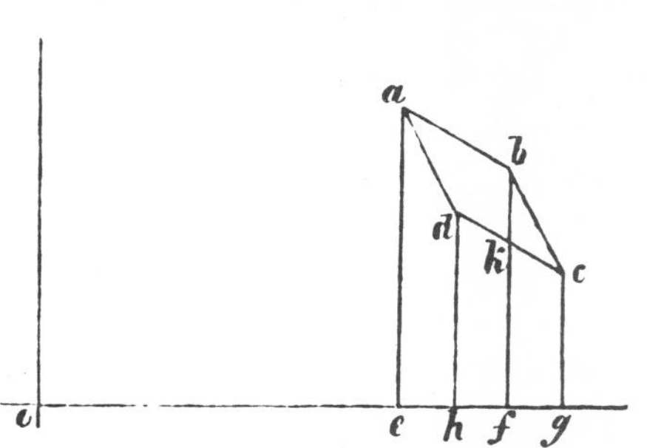

$$dp=\frac{\mathrm{R}dt}{v}.$$

If the increase of volume *ef* be denoted by dv, we obtain the content of the quadrilateral, and with it

$$\textit{The work produced} = \frac{\mathrm{R}\,dv\,dt}{v}. \quad . \quad . \quad . \quad . \quad (1.)$$

We must now determine the quantity of heat consumed during those alterations. Let the amount of heat which must be imparted to change the gas by a definite process from any given state to another, in which its volume is $=v$ and its temperature $=t$, be called Q; and let the changes of volume occurring in the process above described, which are now to be regarded separately, be denoted as follows: *ef* by dv, *hg* by $d'v$, *eh* by δv, and *fg* by $\delta' v$. During an expansion from the volume $oe=v$ to $of=v+dv$, at the constant temperature t, the gas must receive the quantity of heat expressed by

$$\left(\frac{d\mathrm{Q}}{dv}\right)dv;$$

and in accordance with this, during an expansion from $vh=v+\delta v$ to $og=v+\delta v+d'v$ at the temperature $t-dt$, the quantity

$$\left[\frac{d\mathrm{Q}}{dv}+\frac{d}{dv}\left(\frac{d\mathrm{Q}}{dv}\right)\delta v-\frac{d}{dt}\left(\frac{d\mathrm{Q}}{dv}\right)dt\right]d'v.$$

In our case, however, instead of an expansion, a compression has taken place; hence this last expression must be introduced with the negative sign. During the expansion from *of* to *og*, and the compression from *oh* to *oe*, heat has been neither received nor given away; the amount of heat which the gas has received over and above that which it has communicated, or, in other words, ***the quantity of heat consumed***, will therefore be

$$\left(\frac{dQ}{dv}\right)dv - \left[\left(\frac{dQ}{dv}\right) + \frac{d}{dv}\left(\frac{dQ}{dv}\right)\delta v - \frac{d}{dt}\left(\frac{dQ}{dv}\right)dt\right]d'v. \quad (2.)$$

The quantities δv and $d'v$ must now be eliminated; a consideration of the figure furnishes us with the following equation:

$$dv + \delta' v = \delta v + d'v.$$

During its compression from *oh* to *oe*, consequently during its expansion under the same circumstances from *oe* to *oh*, and during the expansion from *of* to *og*, both of which cause a decrease of temperature dt, the gas neither receives nor communicates heat: from this we derive the equations

$$\left(\frac{dQ}{dv}\right)\delta v - \left(\frac{dQ}{dt}\right)dt = 0$$

$$\left[\left(\frac{dQ}{dv}\right) + \frac{d}{dv}\left(\frac{dQ}{dv}\right)dv\right]\delta' v - \left[\left(\frac{dQ}{dt}\right) + \frac{d}{dv}\left(\frac{dQ}{dt}\right)dv\right]dt = 0.$$

From these three equations and equation (2.) the quantities $d'v$, δv and $\delta' v$, may be eliminated; neglecting during the process all differentials of a higher order than the second, we obtain

$$\textit{The heat expended} = \left[\frac{d}{dt}\left(\frac{dQ}{dv}\right) - \frac{d}{dv}\left(\frac{dQ}{dt}\right)\right]dv\,dt. \quad . \quad (3.)$$

Turning now to our maxim, which asserts that the production of a certain quantity of work necessitates the expenditure of a proportionate amount of heat, we may express this in the form of an equation, thus:

$$\frac{\text{The heat expended}}{\text{The work produced}} = A, \quad . \quad . \quad . \quad . \quad (4.)$$

where A denotes a constant which expresses the equivalent of heat for the unit of work. The expressions (1.) and (3.) being introduced into this equation, we obtain

$$\frac{\left[\frac{d}{dt}\left(\frac{dQ}{dv}\right) - \frac{d}{dv}\left(\frac{dQ}{dt}\right)\right]dv\,dt}{\frac{R\,.\,dv\,dt}{v}} = A,$$

or

$$\frac{d}{dt}\left(\frac{dQ}{dv}\right) - \frac{d}{dv}\left(\frac{dQ}{dt}\right) = \frac{A\,.\,R}{v} \quad . \quad . \quad . \quad . \quad . \quad . \quad (\text{II.})$$

This equation may be regarded as the analytical expression of the above maxim applicable to the case of permanent gases. It shows that Q cannot be a function of v and t as long as the two latter are independent of each other. For otherwise, according to the known principle of the differential calculus, that when a function of two variables is differentiated according to both, the order in which this takes place is matter of indifference, the right side of the equation must be equal 0.

The equation can be brought under the form of a *complete differential,* thus:

$$dQ = dU + A.R\frac{a+t}{v}dv, \quad . \quad . \quad . \quad . \qquad \text{(II}a.\text{)}$$

where U denotes an arbitrary function of v and t. This differential equation is of course unintegrable until we find a second condition between the variables, by means of which t may be expressed as a function of v. This is due, however, to the last member alone, and this it is which corresponds to the *exterior* work effected by the alteration; for the differential of this work is pdv, which, when p is eliminated by means of (I.), becomes

$$\frac{R(a+t)}{v}dv.$$

It follows, therefore, in the first place, from (II*a*.), that the entire quantity of heat, Q, absorbed by the gas during a change of volume and temperature may be decomposed into two portions. One of these, U, which comprises the *sensible* heat and the heat necessary for *interior* work, if such be present, fulfils the usual assumption, it is a function of v and t, and is therefore determined by the state of the gas at the beginning and at the end of the alteration; while the other portion, which comprises the heat expended on *exterior* work, depends, not only upon the state of the gas at these two limits, but also upon the manner in which the alterations have been effected throughout. It is shown above that the same conclusion flows directly from the maxim itself.

[*Editor's Note:* Material has been omitted at this point.]

Part II

THE CLASSICAL FORMULATIONS

Editor's Comments on Papers 6 Through 10

6 **THOMSON (Lord Kelvin)**
Excerpt from *On the Dynamical Theory of Heat, with Numerical Results Deduced from Mr. Joule's Equivalent of a Thermal Unit, and M. Regnault's Observations on Steam*

7 **CLAUSIUS**
On the Application of the Theorem of the Equivalence of Transformations to the Internal Work of a Mass of Matter

8 **CLAUSIUS**
On Different Forms of the Fundamental Equations of the Mechanical Theory of Heat and Their Convenience for Application

9 **THOMSON (Lord Kelvin)**
On a Universal Tendency in Nature to the Dissipation of Mechanical Energy

10 **THOMSON (Lord Kelvin)**
On the Economy of the Heating or Cooling of Buildings by Means of Currents of Air

It can be said that the initial two parts of the second law stated in the introduction, as well as their most important consequences, were discovered by Lord Kelvin and Rudolf Clausius. As has often been true in the history of science, it is futile to assign priorities because both of them underwent a lengthy intellectual evolution at about the same time, their views converged, and they influenced each other in the process. Clausius, born in 1822, was barely two years older than Kelvin, and Kelvin's great paper on thermodynamics (Paper 6) appeared in print only one year after Clausius's (Paper 5 of Part I).

Between 1851 (Paper 5 of Part I) and 1867 (Paper 8 of Part II), Clausius published three additional, fundamental papers on thermodynamics, "Abhandlung IV" (1854), "Abhandlung VI" (1862), and "Abhandlung VII" (1863). We reproduce here "Abhandlung VI" (Paper 7) as being the most significant, apart, of course, from the initial, pioneering Paper 5 and the concluding Paper 8, given here in R. B. Lindsay's translation prepared specifically for this volume.[1] Taken together, they allow us to follow in some detail Clausius's route to understanding. From a largely expository paper (Paper 5 of Part I) based on the work of Carnot and Clapeyron, but firmly underpinned with the belief in the conservation of energy and the interchangeability of heat and work, he progresses to an independent formulation of the second law:

> *Heat cannot of itself pass from a colder into a warmer body*

and then to the concept of equivalence of transformations—an unmistakable forerunner of entropy—contained in the statement:

> *If the quantity of heat* Q *of the temperature* t *is produced from work, the equivalent value of this transformation is* Q/T.

The paper further introduces the second part of the second law by the inequality

$$\int \frac{dQ}{T} \geqslant 0 \qquad (1)$$

(note sign convention) quoted as Eq. (Ia). The 1865 paper (reproduced here as Paper 8) marks Clausius's most complete and mature statement of his theory of heat, as does Paper 6 in the case of Kelvin.

In all papers of this part, the principle of energy conservation is accepted, and the second law is based on clear verbal, axiomatic statements:

> **Kelvin:** *It is impossible, by means of inanimate material agency, to derive mechanical effect from any portion of matter by cooling it below the temperature of the coldest of the surrounding objects.*
>
> **Clausius [paraphrased by Kelvin]:** *It is impossible for a self-acting machine, unaided by any external agency, to convey heat from one body to another at a higher temperature.*

[1] According to D. S. Cardwell (*From Watts to Clausius*, Cornell University Press, Ithaca, New York, 1971, p. 327) there exist two complete English translations of Clausius's collected works. These appeared in two volumes (1864 and 1867) as *Abhandlungen über die mechanische Wärmetheorie*, Vieweg, Braunschweig. In addition to the reprints of the original papers, the collected works have been provided by the author with *Supplements*, which he composed later than the papers. The earlier translation by W. R. Browne (1879), which is entitled *The Mechanical Theory of Heat*, does not contain these supplements. The later translation by T. A. Hirst (1887), entitled *The Mechanical Theory of Heat with Its Applications to the Steam Engine and to the Physical Properties of Bodies*, is complete. Both translations are little known and difficult to locate.

It may be interesting to note here that several of Clausius's papers appeared almost simultaneously in German (usually in *Poggendorf's Annalen*), English (usually in the *Philosophical Magazine*) and French (in *Journal de Liouville*).

Modified or copied, these logically equivalent statements have endured for generations. Many critics have quibbled with the semantic side of these formulations, which, it must be admitted, do not live up to our contemporary demands for rigor. Nevertheless, deficient as we may think them to be from this point of view, they constitute a great discovery.

Now, the two laws, the first and the second, have been adopted as a secure basis for the development of the subject, and the well-established properties of reversible Carnot cycles are derived from them instead of the other way around.

Probably the earliest expression for the efficiency of such a cycle is contained in Eq. (8) on p. 190 (**122**) of Paper 6. Using modern notation, it states that

$$\frac{W}{Q} = 1 - \exp\left(-\int_{T_0}^{T} \frac{dT}{T}\right)$$
$$\equiv \frac{T - T_0}{T}, \tag{2}$$

bearing in mind, as already mentioned on p. **14**, that $\mu \equiv 1/T$.

The fact that the Carnot function $C(t) = 1/\mu(t)$ can be taken as the absolute thermodynamic temperature is first and hesitantly stated by Kelvin on p. 199 (**131**) of Paper 6: "It was suggested by Mr. Joule, in a letter dated December 9, 1848, that the true value of μ might be 'inversely as the temperature from zero'; and values for various temperatures calculated by means of the formula

$$\mu = J\,\frac{E}{1 + Et} \tag{3}$$

were given for comparison with those which I had calculated from data regarding steam." Here $E = 0.00366 = 1/273$ is the coefficient of thermal expansion of a (perfect) gas at the ice point. It took a long time, and much unnecessary computational labor, before this "simple" relation became absorbed by science. Clausius's Paper 8 accepts this from the outset.

Compared with the general statements in our introduction, it is seen that these fundamental papers restrict attention to systems without internal deformation variables ζ and are satisfied with a single external deformation variable, $a = v$, so that the only work term considered is

$$dW = p\,dv, \tag{4}$$

as is still the custom in many textbooks. The same papers also clearly distinguish between reversible and irreversible processes but do not associate reversibility with the restriction to equilibrium states. Nevertheless, both authors obviously believe that the principles of thermodynamics apply to other systems, even if they do not state so unequivocally. In contributions (such as Kelvin 1851, 1852; Clausius 1853) not reproduced here, both authors apply thermodynamics to a variety of thermoelectric, thermoelastic, and other phenomena.

The concept of entropy evolved slowly in Clausius's writings; it appears clearly and decisively in Paper 8, which on p. 186 uses the word *Entropie*—derived from the Greek ἡ τροπή (transformation)—for the first time.

Clausius's Paper 8 contains a large number of applications of the first part of the second law to the now-familiar study of relations between the thermodynamic properties of pure substances. These were also known, in a more rudimentary form, to Lord Kelvin (1851, 1852). Clausius's paper, naturally, repeats the statement of the second part of the second law as well as the equivalent of the concept of entropy production Θ (denoted here by N in Eq. (71) on p. **191**) created in an irreversible process. When stating the second part of the second law, Clausius is careful to restrict it to the integral forms.

$$\int \frac{dQ}{T} \leqslant 0 \tag{5}$$

and

$$N = S - S_0 - \int \frac{dQ}{T}. \tag{6}$$

Since the difference $S - S_0$ is calculated along a reversible path, it follows—even though Clausius does not dot the *i*s in this respect—that the integrals must refer to initial and final states that are equilibrium states. The intermediate states may be nonequilibrium states because Eq. (6) is applicable to irreversible processes with $N > 0$. This Clausius recognizes quite explicitly when he states that the first law,

$$Q = U - U_0 + w, \tag{7}$$

applies to reversible as well as irreversible processes, the difference between them consisting in the fact that the work w is different in each.

The inequality

$$dS \geqslant \frac{dQ}{T} \qquad (8)$$

appeared, without comment, in Clausius's later writings,[2] but as we shall see in Part IV, its validity should be questioned.

Clausius's Paper 8 ends with the famous statements about the "universe":

1. *The energy of the universe is constant.*
2. *The entropy of the universe tends to a maximum.*

Sommerfeld (1956) and others have taken exception to these unnecessary (and unwarranted) generalizations because it is difficult to see that the "universe" necessarily constitutes a closed and isolated thermodynamic system. Moreover, to make any statement about the universe's entropy, it would be necessary to observe it in equilibrium states, which is hardly imaginable.

Nevertheless, in a historical recapitulation it is not difficult to excuse the great thermodynamicist, who was then forty-three years old, for letting himself be carried away by his enthusiasm. With a certain amount of good will it is even possible to interpret the second sentence above as a statement of the equilibrium principle.

In addition to the seminal papers of Kelvin and Clausius, I include two further papers by Kelvin. The first one, Paper 9, contains an essay on dissipation and maximum work, whereas Paper 10 establishes Kelvin's priority as the inventor of the heat pump. Digressing somewhat, I cannot restrain myself from drawing attention to the fact that the first practical application of the heat pump was achieved around 1938, 86 years after Lord Kelvin had clearly ennunciated its principle!

[2] In 1876 Clausius published *Die mechanische Wärmetheorie* which was the outgrowth of a detailed revision of an earlier edition of his two-volume collected works entitled *Abhandlungen über die mechanische Wärmetheorie.* The earlier work, consisting of Clausius's original papers and extensive commentaries, does not contain the above inequality (8). It appears in the form $dQ \leqslant T\,dS$ on p. 244 of the later work as an "obvious" consequence of

$$\int \frac{dQ}{T} \leqslant 0$$

for irreversible processes and of $dQ = T\,dS$ for reversible processes.

REFERENCES

Clausius, R. 1853. Über die Anwendung der mechanischen Wärmetheorie auf die thermoelectrischen Erscheinungen. *Pogg. Ann.*, **40**:513. Also *Abhandlungen über die mechanische Wärmetheorie, Vol. II*, pp. 175–201. (See footnote on p. **104**).

Sommerfeld, A. 1956. *Lectures in Theoretical Physics, Vol. V: Thermodynamics and Statistical Mechanics*, J. Kestin, trans., Academic Press, New York, p. 38.

Thomson, W. (Lord Kelvin). 1851, 1852. On the Dynamical Theory of Heat, with Numerical Results Deduced from Mr. Joule's Equivalent of a Thermal Unit, and M. Regnault's Observations on Steam. *Trans. Roy. Soc. Edinb.* (1851) and *Phil. Mag.* 4 (1852). Also Art. XLVII, Part III of which is entitled "Applications of the Dynamical Theory To Establish Relations between the Physical Properties of All Substances" (especially Part VI on pp. 232–291 and Part VII on pp. 291–316) of *Mathematical and Physical Papers of William Thomson*, Vol. I, Cambridge University Press, Cambridge, 1882, pp. 174–200.

6

Reprinted from pp. 174–200 of *Mathematical and Physical Papers of William Thomson*, Vol. 1, Cambridge University Press, 1882, 571 pp.

ON THE DYNAMICAL THEORY OF HEAT, WITH NUMERICAL RESULTS DEDUCED FROM MR. JOULE'S EQUIVALENT OF A THERMAL UNIT, AND M. REGNAULT'S OBSERVATIONS ON STEAM

W. Thomson

Introductory Notice.

1. SIR HUMPHRY DAVY, by his experiment of melting two pieces of ice by rubbing them together, established the following proposition:—"The phenomena of repulsion are not dependent on a peculiar elastic fluid for their existence, or caloric does not exist." And he concludes that heat consists of a motion excited among the particles of bodies. "To distinguish this motion from others, and to signify the cause of our sensation of heat," and of the expansion or expansive pressure produced in matter by heat, "the name *repulsive* motion has been adopted *."

2. The dynamical theory of heat, thus established by Sir Humphry Davy, is extended to radiant heat by the discovery of phenomena, especially those of the polarization of radiant heat, which render it excessively probable that heat propagated through "vacant space," or through diathermanic substances, consists of waves of transverse vibrations in an all-pervading medium.

* From Davy's first work, entitled *An Essay on Heat, Light, and the Combinations of Light*, published in 1799, in "Contributions to Physical and Medical Knowledge, principally from the West of England, collected by Thomas Beddoes, M.D.," and republished in Dr Davy's edition of his brother's collected works, Vol. II. Lond. 1836.

3. The recent discoveries made by Mayer and Joule*, of the generation of heat through the friction of fluids in motion, and by the magneto-electric excitation of galvanic currents, would either of them be sufficient to demonstrate the immateriality of heat; and would so afford, if required, a perfect confirmation of Sir Humphry Davy's views.

4. Considering it as thus established, that heat is not a substance, but a dynamical form of mechanical effect, we perceive that there must be an equivalence between mechanical work and heat, as between cause and effect. The first published statement of this principle appears to be in Mayer's *Bemerkungen über die Kräfte der unbelebten Natur*†, which contains some correct views regarding the mutual convertibility of heat and mechanical effect, along with a false analogy between the approach of a weight to the earth and a diminution of the volume of a continuous substance, on which an attempt is founded to find numerically the mechanical equivalent of a given quantity of heat. In a paper published about fourteen months later, "On the Calorific Effects of Magneto-Electricity and the Mechanical Value of Heat‡," Mr Joule of Manchester expresses very distinctly the consequences regarding the mutual convertibility of heat and mechanical effect which follow from the fact, that heat is not a substance but a state of motion; and investigates on unquestionable principles the "absolute numerical relations," according to which heat is connected with mechanical power; verifying experimentally, that whenever heat is generated from purely mechanical action, and no other effect produced, whether it be by means of the friction of fluids or by the magneto-electric excitation of galvanic currents, the same quantity is generated by the same amount of work spent; and determining the actual amount of work, in foot-pounds,

* In May, 1842, Mayer announced in the *Annalen* of Wöhler and Liebig, that he had raised the temperature of water from 12° to 13° Cent. by agitating it. In August, 1843, Joule announced to the British Association "That heat is evolved by the passage of water through narrow tubes;" and that he had "obtained one degree of heat per lb. of water from a mechanical force capable of raising 770 lbs. to the height of one foot;" and that heat is generated when work is spent in turning a magneto-electric machine, or an electro-magnetic engine. (See his paper "On the Calorific Effects of Magneto-Electricity, and on the Mechanical Value of Heat."—*Phil. Mag.*, Vol. XXIII., 1843.)

† *Annalen* of Wöhler and Liebig, May, 1842.

‡ British Association, August, 1843; and *Phil. Mag.*, Sept., 1843.

required to generate a unit of heat, which he calls "the mechanical equivalent of heat." Since the publication of that paper, Mr Joule has made numerous series of experiments for determining with as much accuracy as possible the mechanical equivalent of heat so defined, and has given accounts of them in various communications to the British Association, to the *Philosophical Magazine*, to the Royal Society, and to the French Institute.

5. Important contributions to the dynamical theory of heat have recently been made by Rankine and Clausius; who, by mathematical reasoning analogous to Carnot's on the motive power of heat, but founded on an axiom contrary to his fundamental axiom, have arrived at some remarkable conclusions. The researches of these authors have been published in the *Transactions* of this Society, and in Poggendorff's *Annalen*, during the past year; and they are more particularly referred to below in connexion with corresponding parts of the investigations at present laid before the Royal Society.

[Various statements regarding animal heat, and the heat of combustion and chemical combination, are made in the writings of Liebig (as, for instance, the statement quoted in the foot-note added to § 18 below), which virtually imply the convertibility of heat into mechanical effect, and which are inconsistent with any other than the dynamical theory of heat.]

6. The object of the present paper is threefold:—

(1) To show what modifications of the conclusions arrived at by Carnot, and by others who have followed his peculiar mode of reasoning regarding the motive power of heat, must be made when the hypothesis of the dynamical theory, contrary as it is to Carnot's fundamental hypothesis, is adopted.

(2) To point out the significance in the dynamical theory, of the numerical results deduced from Regnault's observations on steam, and communicated about two years ago to the Society, with an account of Carnot's theory, by the author of the present paper; and to show that by taking these numbers (subject to correction when accurate experimental data regarding the density of saturated steam shall have been afforded), in connexion with Joule's mechanical equivalent of a thermal unit, a complete theory

of the motive power of heat, within the temperature limits of the experimental data, is obtained.

(3) To point out some remarkable relations connecting the physical properties of all substances, established by reasoning analogous to that of Carnot, but founded in part on the contrary principle of the dynamical theory.

PART I.

Fundamental Principles in the Theory of the Motive Power of Heat.

7. According to an obvious principle, first introduced, however, into the theory of the motive power of heat by Carnot, mechanical effect produced in any process cannot be said to have been derived from a purely thermal source, unless at the end of the process all the materials used are in precisely the same physical and mechanical circumstances as they were at the beginning. In some conceivable "thermo-dynamic engines," as for instance Faraday's floating magnet, or Barlow's "wheel and axle," made to rotate and perform work uniformly by means of a current continuously excited by heat communicated to two metals in contact, or the thermo-electric rotatory apparatus devised by Marsh, which has been actually constructed; this condition is fulfilled at every instant. On the other hand, in all thermo-dynamic engines, founded on electrical agency, in which discontinuous galvanic currents, or pieces of soft iron in a variable state of magnetization, are used, and in all engines founded on the alternate expansions and contractions of media, there are really alterations in the condition of materials; but, in accordance with the principle stated above, these alterations must be strictly periodical. In any such engine, the series of motions performed during a period, at the end of which the materials are restored to precisely the same condition as that in which they existed at the beginning, constitutes what will be called a complete cycle of its operations. Whenever in what follows, *the work done* or *the mechanical effect produced* by a thermo-dynamic engine is mentioned without qualification, it must be understood that the mechanical effect produced, either in a non-varying engine, or in a complete cycle, or any number of complete cycles of a periodical engine, is meant.

8. The *source of heat* will always be supposed to be a hot body at a given constant temperature, put in contact with some part of the engine; and when any part of the engine is to be kept from rising in temperature (which can only be done by drawing off whatever heat is deposited in it), this will be supposed to be done by putting a cold body, which will be called the refrigerator, at a given constant temperature in contact with it.

9. The whole theory of the motive power of heat is founded on the two following propositions, due respectively to Joule, and to Carnot and Clausius.

PROP. I. (Joule).—When equal quantities of mechanical effect are produced by any means whatever from purely thermal sources, or lost in purely thermal effects, equal quantities of heat are put out of existence or are generated.

PROP. II. (Carnot and Clausius).—If an engine be such that, when it is worked backwards, the physical and mechanical agencies in every part of its motions are all reversed, it produces as much mechanical effect as can be produced by any thermodynamic engine, with the same temperatures of source and refrigerator, from a given quantity of heat.

10. The former proposition is shown to be included in the general "principle of mechanical effect," and is so established beyond all doubt by the following demonstration.

11. By whatever direct effect the heat gained or lost by a body in any conceivable circumstances is tested, the measurement of its quantity may always be founded on a determination of the quantity of some standard substance, which it or any equal quantity of heat could raise from one standard temperature to another; the test of equality between two quantities of heat being their capability of raising equal quantities of any substance from any temperature to the same higher temperature. Now, according to the dynamical theory of heat, the temperature of a substance can only be raised by working upon it in some way so as to produce increased thermal motions within it, besides effecting any modifications in the mutual distances or arrangements of its particles which may accompany a change of temperature. The work necessary to produce this total mechanical effect is of course proportional to the quantity of the substance raised from one

standard temperature to another; and therefore when a body, or a group of bodies, or a machine, parts with or receives heat, there is in reality mechanical effect produced from it, or taken into it, to an extent precisely proportional to the quantity of heat which it emits or absorbs. But the work which any external forces do upon it, the work done by its own molecular forces, and the amount by which the half *vis viva* of the thermal motions of all its parts is diminished, must together be equal to the mechanical effect produced from it; and consequently, to the mechanical equivalent of the heat which it emits (which will be positive or negative, according as the sum of those terms is positive or negative). Now let there be either no molecular change or alteration of temperature in any part of the body, or, by a cycle of operations, let the temperature and physical condition be restored exactly to what they were at the beginning; the second and third of the three parts of the work which it has to produce vanish; and we conclude that the heat which it emits or absorbs will be the thermal equivalent of the work done upon it by external forces, or done by it against external forces; which is the proposition to be proved.

12. The demonstration of the second proposition is founded on the following axiom:—

It is impossible, by means of inanimate material agency, to derive mechanical effect from any portion of matter by cooling it below the temperature of the coldest of the surrounding objects.*

13. To demonstrate the second proposition, let A and B be two thermo-dynamic engines, of which B satisfies the conditions expressed in the enunciation; and let, if possible, A derive more work from a given quantity of heat than B, when their sources and refrigerators are at the same temperatures, respectively. Then on account of the condition of complete *reversibility* in all its operations which it fulfils, B may be worked backwards, and made to restore any quantity of heat to its source, by the expenditure of the amount of work which, by its forward action, it would derive from the same quantity of heat. If, therefore, B be

* If this axiom be denied for all temperatures, it would have to be admitted that a self-acting machine might be set to work and produce mechanical effect by cooling the sea or earth, with no limit but the total loss of heat from the earth and sea, or, in reality, from the whole material world.

worked backwards, and made to restore to the source of A (which we may suppose to be adjustable to the engine B) as much heat as has been drawn from it during a certain period of the working of A, a smaller amount of work will be spent thus than was gained by the working of A. Hence, if such a series of operations of A forwards and of B backwards be continued, either alternately or simultaneously, there will result a continued production of work without any continued abstraction of heat from the source; and, by Prop. I., it follows that there must be more heat abstracted from the refrigerator by the working of B backwards than is deposited in it by A. Now it is obvious that A might be made to spend part of its work in working B backwards, and the whole might be made self-acting. Also, there being no heat either taken from or given to the source on the whole, all the surrounding bodies and space except the refrigerator might, without interfering with any of the conditions which have been assumed, be made of the same temperature as the source, whatever that may be. We should thus have a self-acting machine, capable of drawing heat constantly from a body surrounded by others at a higher temperature, and converting it into mechanical effect. But this is contrary to the axiom, and therefore we conclude that the hypothesis that A derives more mechanical effect from the same quantity of heat drawn from the source than B, is false. Hence no engine whatever, with source and refrigerator at the same temperatures, can get more work from a given quantity of heat introduced than any engine which satisfies the condition of reversibility, which was to be proved.

14. This proposition was first enunciated by Carnot, being the expression of his criterion of a perfect thermo-dynamic engine*. He proved it by demonstrating that a negation of it would require the admission that there might be a self-acting machine constructed which would produce mechanical effect indefinitely, without any source either in heat or the consumption of materials, or any other physical agency; but this demonstration involves, fundamentally, the assumption that, in "a complete cycle of operations," the medium parts with exactly the same quantity of heat as it receives. A very strong expression of doubt regarding the truth of this assumption, as a universal principle, is given by

* Account of Carnot's *Theory*, § 13.

Carnot himself*; and that it is false, where mechanical work is, on he whole, either gained or spent in the operations, may (as I have tried to show above) be considered to be perfectly certain. It must then be admitted that Carnot's original demonstration utterly fails, but we cannot infer that the proposition concluded is false. The truth of the conclusion appeared to me, indeed, so probable, that I took it in connexion with Joule's principle, on account of which Carnot's demonstration of it fails, as the foundation of an investigation of the motive power of heat in air-engines or steam-engines through finite ranges of temperature, and obtained about a year ago results, of which the substance is given in the second part of the paper at present communicated to the Royal Society. It was not until the commencement of the present year that I found the demonstration given above, by which the truth of the proposition is established upon an axiom (§ 12) which I think will be generally admitted. It is with no wish to claim priority that I make these statements, as the merit of first establishing the proposition upon correct principles is entirely due to Clausius, who published his demonstration of it in the month of May last year, in the second part of his paper on the motive power of heat†. I may be allowed to add, that I have given the demonstration exactly as it occurred to me before I knew that Clausius had either enunciated or demonstrated the proposition. The following is the axiom on which Clausius' demonstration is founded :—

It is impossible for a self-acting machine, unaided by any external agency, to convey heat from one body to another at a higher temperature.

It is easily shown, that, although this and the axiom I have used are different in form, either is a consequence of the other. The reasoning in each demonstration is strictly analogous to that which Carnot orginally gave.

15. A complete theory of the motive power of heat would consist of the application of the two propositions demonstrated above, to every possible method of producing mechanical effect from thermal agency‡. As yet this has not been done for the

* Account of Carnot's *Theory*, § 6.

† Poggendorff's *Annalen*, referred to above.

‡ "There are at present known two, and only two, distinct ways in which

electrical method, as far as regards the criterion of a perfect engine implied in the second proposition, and probably cannot be done without certain limitations; but the application of the first proposition has been very thoroughly investigated, and verified experimentally by Mr Joule in his researches "On the Calorific Effects of Magneto-Electricity;" and on it is founded one of his ways of determining experimentally the mechanical equivalent of heat. Thus, from his discovery of the laws of generation of heat in the galvanic circuit*, it follows that when mechanical work by means of a magneto-electric machine is the source of the galvanism, the heat generated in any given portion of the fixed part of the circuit is proportional to the whole work spent; and from his experimental demonstration that heat is developed in any moving part of the circuit at exactly the same rate as if it were at rest, and traversed by a current of the same strength, he is enabled to conclude—

(1) That heat may be created by working a magneto-electric machine.

(2) That if the current excited be not allowed to produce any other than thermal effects, the total quantity of heat produced is in all circumstances exactly proportional to the quantity of work spent.

16. Again, the admirable discovery of Peltier, that cold is produced by an electrical current passing from bismuth to antimony, is referred to by Joule†, as showing how it may be proved

mechanical effect can be obtained from heat. One of these is by the alterations of volume which bodies experience through the action of heat; the other is through the medium of electric agency."—"Account of Carnot's Theory," § 4. (*Transactions*, Vol. XVI. part 5.)

* That, in a given fixed part of the circuit, the heat evolved in a given time is proportional to the square of the strength of the current, and for different fixed parts, with the same strength of current, the quantities of heat evolved in equal times are as the resistances. A paper by Mr Joule, containing demonstrations of these laws, and of others on the relations of the chemical and thermal agencies concerned, was communicated to the Royal Society on the 17th of December, 1840, but was not published in the *Transactions*. (See abstract containing a statement of the laws quoted above, in the *Philosophical Magazine*, Vol. XVIII. p. 308.) It was published in the *Philosophical Magazine* in October, 1841 (Vol. XIX. p. 260).

† [Note of March 20, 1852, added in *Phil. Mag.* reprint. In the introduction to his paper on the "Calorific Effects of Magneto-Electricity," &c., *Phil. Mag.*, 1843. I take this opportunity of mentioning that I have only recently become ac-

that, when an electrical current is continuously produced from a purely thermal source, the quantities of heat evolved electrically in the different homogeneous parts of the circuit are only compensations for a loss from the junctions of the different metals, or that, when the effect of the current is entirely thermal, there must be just as much heat emitted from the parts not affected by the source as is taken from the source.

17. Lastly*, when a current produced by thermal agency is made to work an engine and produce mechanical effect, there will be less heat emitted from the parts of the circuit not affected by the source than is taken in from the source, by an amount precisely equivalent to the mechanical effect produced; since Joule demonstrates experimentally, that a current from any kind of

quainted with Helmholtz's admirable treatise on the principle of mechanical effect (*Ueber die Erhaltung der Kraft*, von Dr H. Helmholtz. Berlin. G. Reimer, 1847), having seen it for the first time on the 20th of January of this year; and that I should have had occasion to refer to it on this, and on numerous other points of the dynamical theory of heat, the mechanical theory of electrolysis, the theory of electro-magnetic induction, and the mechanical theory of thermo-electric currents, in various papers communicated to the Royal Society of Edinburgh, and to this Magazine, had I been acquainted with it in time.—W. T., March 20, 1852.]

* This reasoning was suggested to me by the following passage contained in a letter which I received from Mr Joule on the 8th of July, 1847. "In Peltier's experiment on cold produced at the bismuth and antimony solder, we have an instance of the conversion of heat into the mechanical force of the current," which must have been meant as an answer to a remark I had made, that no evidence could be adduced to show that heat is ever put out of existence. I now fully admit the force of that answer; but it would require a proof that there is more heat put out of existence at the heated soldering [or in this and other parts of the circuit] than is created at the cold soldering [and the remainder of the circuit, when a machine is driven by the current] to make the "evidence" be *experimental*. That this is the case I think is certain, because the statements of § 16 in the text are demonstrated consequences of the first fundamental proposition; but it is still to be remarked, that neither in this nor in any other case of the production of mechanical effect from purely thermal agency, has the ceasing to exist of an equivalent quantity of heat been demonstrated otherwise than theoretically. It would be a very great step in the experimental illustration (or *verification*, for those who consider such to be necessary) of the dynamical theory of heat, to actually show in any one case a loss of heat; and it might be done by operating through a very considerable range of temperatures with a good air-engine or steam-engine, not allowed to waste its work in friction. As will be seen in Part. II. of this paper, no experiment of any kind could show a considerable loss of heat without employing bodies differing considerably in temperature; for instance, a loss of as much as ·098, or about one-tenth of the whole heat used, if the temperature of all the bodies used be between 0^0 and 30^0 Cent.

source driving an engine, produces in the engine just as much less heat than it would produce in a fixed wire exercising the same resistance as is equivalent to the mechanical effect produced by the engine.

18. The quality of thermal effects, resulting from equal causes through very different means, is beautifully illustrated by the following statement, drawn from Mr Joule's paper on magneto-electricity*.

Let there be three equal and similar galvanic batteries furnished with equal and similar electrodes; let A_1 and B_1 be the terminations of the electrodes (or wires connected with the two poles) of the first battery, A_2 and B_2 the terminations of the corresponding electrodes of the second, and A_3 and B_3 of the third battery. Let A_1 and B_1 be connected with the extremities of a long fixed wire; let A_2 and B_2 be connected with the "poles" of an electrolytic apparatus for the decomposition of water; and let A_3 and B_3 be connected with the *poles* (or *ports* as they might be called) of an electro-magnetic engine. Then if the length of the wire between A_1 and B_1, and the speed of the engine between A_3 and B_3, be so adjusted that the strength of the current (which for simplicity we may suppose to be continuous and perfectly uniform in each case) may be the same in the three circuits, there will be more heat given out in any time in the wire between A_1 and B_1 than in the electrolytic apparatus between A_2 and B_2, or the working engine between A_3 and B_3. But if the hydrogen were allowed to burn in the oxygen, within the electrolytic vessel, and the engine to waste all its work without producing any other than thermal effects (as it would do, for instance, if all its work were spent in continuously agitating a limited fluid mass), the total heat emitted would be precisely the same in each of these two pieces of apparatus as in the wire between A_1 and B_1. It is worthy of remark that these propositions are *rigorously* true, being demonstrable consequences of the fundamental principle of the dynamical theory of heat, which have been discovered by Joule,

* In this paper reference is made to his previous paper "On the Heat of Electrolysis" (published in Vol. VII. part 2, of the second series of the Literary and Philosophical Society of Manchester) for experimental demonstration of those parts of the theory in which chemical action is concerned.

and illustrated and verified most copiously in his experimental researches*.

19. Both the fundamental propositions may be applied in a perfectly rigorous manner to the second of the known methods of producing mechanical effect from thermal agency. This application of the first of the two fundamental propositions has already been published by Rankine and Clausius; and that of the second, as Clausius showed in his published paper, is simply Carnot's unmodified investigation of the relation between the mechanical effect produced and the thermal circumstances from which it originates, in the case of an expansive engine working within an infinitely small range of temperatures. The simplest investigation of the consequences of the first proposition in this application, which has occurred to me, is the following, being merely the modification of an analytical expression of Carnot's axiom regarding the permanence of heat, which was given in my former paper†, required to make it express, not Carnot's axiom, but Joule's.

20. Let us suppose a mass‡ of any substance, occupying a volume v, under a pressure p uniform in all directions, and at a temperature t, to expand in volume to $v + dv$, and to rise in tem-

[* Note of March 20, 1852, added in *Phil. Mag.* reprint. I have recently met with the following passage in Liebig's *Animal Chemistry* (3rd edit. London, 1846, p. 43), in which the dynamical theory of the heat both of combustion and of the galvanic battery is indicated, if not fully expressed:—"When we kindle a fire under a steam-engine, and employ the power obtained to produce heat by friction, it is impossible that the heat thus obtained can ever be greater than that which was required to heat the boiler; and if we use the galvanic current to produce heat, the amount of heat obtained is never in any circumstances greater than we might have by the combustion of the zinc which has been dissolved in the acid."

A paper "On the Heat of Chemical Combination," by Dr Thomas Woods, published last October in the *Philosophical Magazine*, contains an independent and direct experimental demonstration of the proposition stated in the text regarding the comparative thermal effects in a fixed metallic wire, and an electrolytic vessel for the decomposition of water, produced by a galvanic current.—W. T., March 20, 1852.]

† "Account of Carnot's Theory," foot-note on § 26.

‡ This may have parts consisting of different substances, or of the same substance in different states, provided the temperature of all be the same. See below Part III., § 53—56.

perature to $t + dt$. The quantity of work which it will produce will be

$$pdv;$$

and the quantity of heat which must be added to it to make its temperature rise during the expansion to $t + dt$ may be denoted by

$$Mdv + Ndt.$$

The mechanical equivalent of this is

$$J(Mdv + Ndt),$$

if J denote the mechanical equivalent of a unit of heat. Hence the mechanical measure of the total external effect produced in the circumstances is

$$(p - JM)\,dv - JNdt.$$

The total external effect, after any finite amount of expansion, accompanied by any continuous change of temperature, has taken place, will consequently be, in mechanical terms,

$$\int \{(p - JM)\,dv - JNdt\};$$

where we must suppose t to vary with v, so as to be the actual temperature of the medium at each instant, and the integration with reference to v must be performed between limits corresponding to the initial and final volumes. Now if, at any subsequent time, the volume and temperature of the medium become what they were at the beginning, however arbitrarily they may have been made to vary in the period, the total external effect must, according to Prop. I., amount to nothing; and hence

$$(p - JM)\,dv - JNdt^*$$

must be the differential of a function of two independent variables, or we must have

$$\frac{d(p - JM)}{dt} = \frac{d(-JN)}{dv} \quad\text{.................... (1)},$$

this being merely the analytical expression of the condition, that the preceding integral may vanish in every case in which the

[* The integral function $\int \{(JM - p)\,dv + JNdt\}$ may obviously be called the *mechanical energy* of the fluid mass; as (when the constant of integration is properly assigned) it expresses the whole work the fluid has in it to produce. The consideration of this function is the subject of a short paper communicated to the Royal Society of Edinburgh, Dec. 15, 1851, as an appendix to the paper at present republished; (see below Part v. §§ 81—96).]

initial and final values of v and t are the same, respectively. Observing that J is an absolute constant, we may put the result into the form

$$\frac{dp}{dt} = J\left(\frac{dM}{dt} - \frac{dN}{dv}\right) \quad \text{..................(2).}$$

This equation expresses, in a perfectly comprehensive manner, the application of the first fundamental proposition to the thermal and mechanical circumstances of any substance whatever, under uniform pressure in all directions, when subjected to any possible variations of temperature, volume and pressure.

21. The corresponding application of the second fundamental proposition is completely expressed by the equation

$$\frac{dp}{dt} = \mu M \quad \text{..........................(3),}$$

where μ denotes what is called "Carnot's function," a quantity which has an absolute value, the same for all substances for any given temperature, but which may vary with the temperature in a manner that can only be determined by experiment. To prove this proposition, it may be remarked in the first place that Prop. II. could not be true for every case in which the temperature of the refrigerator differs infinitely little from that of the source, without being true universally. Now, if a substance be allowed first to expand from v to $v + dv$, its temperature being kept constantly t; if, secondly, it be allowed to expand further, without either emitting or absorbing heat till its temperature goes down through an infinitely small range, to $t - \tau$; if, thirdly, it be compressed at the constant temperature $t - \tau$, so much (actually by an amount differing from dv by only an infinitely small quantity of the second order), that when, fourthly, the volume is further diminished to v without the medium's being allowed to either emit or absorb heat, its temperature may be exactly t; it may be considered as constituting a thermo-dynamic engine which fulfils Carnot's condition of complete reversibility. Hence, by Prop. II., it must produce the same amount of work for the same quantity of heat absorbed in the first operation, as any other substance similarly operated upon through the same range of temperatures. But $\frac{dp}{dt}\tau . dv$ is obviously the whole work done in the complete cycle, and (by the definition of M in § 20) Mdv is the quantity of heat absorbed in the first operation. Hence the value of

$$\frac{\frac{dp}{dt}\tau . dv}{Mdv}, \quad \text{or} \quad \frac{\frac{dp}{dt}}{M}\tau,$$

must be the same for all substances, with the same values of t and τ; or, since τ is not involved except as a factor, we must have

$$\frac{\frac{dp}{dt}}{M} = \mu \ldots\ldots\ldots\ldots\ldots\ldots\ldots\ldots\ldots (4),$$

where μ depends only on t; from which we conclude the proposition which was to be proved.

[Note of Nov. 9, 1881. Elimination of $\frac{dp}{dt}$ by (2) from (4) gives

$$\frac{J\left(\frac{dM}{dt} - \frac{dN}{dv}\right)}{M} = \mu \ldots\ldots\ldots\ldots\ldots\ldots (4'),$$

a very convenient and important formula.]

22. The very remarkable theorem that $\frac{\frac{dp}{dt}}{M}$ must be the same for all substances at the same temperature, was first given (although not in precisely the same terms) by Carnot, and demonstrated by him, according to the principles he adopted. We have now seen that its truth may be satisfactorily established without adopting the false part of his principles. Hence all Carnot's conclusions, and all conclusions derived by others from his theory, which depend merely on equation (3), require no modification when the dynamical theory is adopted. Thus, all the conclusions contained in Sections I., II., and III., of the Appendix to my "Account of Carnot's Theory" [Art. XLI. §§ 43—53 above], and in the paper immediately following it in the *Transactions* [and in the present reprint], entitled "Theoretical Considerations on the Effect of Pressure in Lowering the Freezing Point of Water," by my elder brother, still hold. Also, we see that Carnot's expression for the mechanical effect derivable from a given quantity of heat by means of a perfect engine in which the range of temperatures is infinitely small, expresses truly the greatest effect

which can possibly be obtained in the circumstances; although it is in reality only an infinitely small fraction of the whole mechanical equivalent of the heat supplied; the remainder being irrecoverably lost to man, and therefore "wasted," although not *annihilated.*

23. On the other hand, the expression for the mechanical effect obtainable from a given quantity of heat entering an engine from a "source" at a given temperature, when the range down to the temperature of the cold part of the engine or the "refrigerator" is finite, will differ most materially from that of Carnot; since, a finite quantity of mechanical effect being now obtained from a finite quantity of heat entering the engine, a finite fraction of this quantity must be converted from heat into mechanical effect. The investigation of this expression, with numerical determinations founded on the numbers deduced from Regnault's observations on steam, which are shown in Tables I. and II. of my former paper, constitutes the second part of the paper at present communicated.

PART II.

On the Motive Power of Heat through Finite Ranges of Temperature.

24. It is required to determine the quantity of work which a perfect engine, supplied from a source at any temperature, S, and parting with its waste heat to a refrigerator at any lower temperature, T, will produce from a given quantity, H, of heat drawn from the source.

25. We may suppose the engine to consist of an infinite number of perfect engines, each working within an infinitely small range of temperature, and arranged in a series of which the source of the first is the given source, the refrigerator of the last the given refrigerator, and the refrigerator of each intermediate engine is the source of that which follows it in the series. Each of these engines will, in any time, emit just as much less heat to its refrigerator than is supplied to it from its source, as is the equivalent of the mechanical work which it produces. Hence if t and $t+dt$ denote respectively the temperatures of the refrigerator and

source of one of the intermediate engines, and if q denote the quantity of heat which this engine discharges into its refrigerator in any time, and $q+dq$ the quantity which it draws from its source in the same time, the quantity of work which it produces in that time will be Jdq according to Prop. I., and it will also be $q\mu dt$ according to the expression of Prop. II., investigated in § 21; and therefore we must have

$$Jdq = q\mu dt.$$

Hence, supposing that the quantity of heat supplied from the first source, in the time considered is H, we find by integration

$$\log \frac{H}{q} = \frac{1}{J}\int_t^S \mu dt.$$

But the value of q, when $t = T$, is the final remainder discharged into the refrigerator at the temperature T; and therefore, if this be denoted by R, we have

$$\log \frac{H}{R} = \frac{1}{J}\int_T^S \mu dt \ldots\ldots\ldots\ldots\ldots\ldots\ldots\ldots(5);$$

from which we deduce

$$R = H\epsilon^{-\frac{1}{J}\int_T^S \mu dt} \ldots\ldots\ldots\ldots\ldots\ldots\ldots\ldots(6).$$

Now the whole amount of work produced will be the mechanical equivalent of the quantity of heat lost; and, therefore, if this be denoted by W, we have

$$W = J(H - R) \ldots\ldots\ldots\ldots\ldots\ldots\ldots\ldots(7),$$

and consequently, by (6),

$$W = JH\{1 - \epsilon^{-\frac{1}{J}\int_T^S \mu dt}\} \ldots\ldots\ldots\ldots\ldots\ldots(8).$$

26. To compare this with the expression $H\int_T^S \mu dt$, for the duty indicated by Carnot's theory*, we may expand the exponential in the preceding equation, by the usual series. We thus find

$$\left.\begin{aligned} W &= \left(1 - \frac{\theta}{1.2} + \frac{\theta^2}{1.2.3} - \&c.\right).H\int_T^S \mu dt \\ \text{where}\quad \theta &= \frac{1}{J}\int_T^S \mu dt \end{aligned}\right\} \ldots\ldots(9),$$

* "Account," &c., Equation 7, § 31. [Art. XLI. above.]

This shows that the work really produced, which always falls short of the duty indicated by Carnot's theory, approaches more and more nearly to it as the range is diminished; and ultimately, when the range is infinitely small, is the same as if Carnot's theory required no modification, which agrees with the conclusion stated above in § 22.

27. Again, equation (8) shows that the real duty of a given quantity of heat supplied from the source increases with every increase of the range; but that instead of increasing indefinitely in proportion to $\int_T^S \mu dt$, as Carnot's theory makes it do, it never reaches the value JH, but approximates to this limit, as $\int_T^S \mu dt$ is increased without limit. Hence Carnot's remark* regarding the practical advantage that may be anticipated from the use of the air-engine, or from any method by which the range of temperatures may be increased, loses only a part of its importance, while a much more satisfactory view than his of the practical problem is afforded. Thus we see that, although the full equivalent of mechanical effect cannot be obtained even by means of a perfect engine, yet when the actual source of heat is at a high enough temperature above the surrounding objects, we may get more and more nearly the whole of the admitted heat converted into mechanical effect, by simply increasing the effective range of temperature in the engine.

28. The preceding investigation (§ 25) shows that the value of Carnot's function, μ, for all temperatures within the range of the engine, and the absolute value of Joule's equivalent, J, are enough of data to calculate the amount of mechanical effect of a perfect engine of any kind, whether a steam-engine, an air-engine, or even a thermo-electric engine; since, according to the axiom stated in § 12, and the demonstration of Prop. II., no inanimate material agency could produce more mechanical effect from a given quantity of heat, with a given available range of temperatures, than an engine satisfying the criterion stated in the enunciation of the proposition.

* "Account," &c. Appendix, Section IV. [Art. XLI. above.]

29. The mechanical equivalent of a thermal unit Fahrenheit, or the quantity of heat necessary to raise the temperature of a pound of water from 32° to 33° Fahr., has been determined by Joule in foot-pounds at Manchester, and the value which he gives as his best determination is 772·69. Mr Rankine takes, as the result of Joule's determination 772, which he estimates must be within $\frac{1}{300}$ of its own amount, of the truth. If we take $772\frac{2}{9}$ as the number, we find, by multiplying it by $\frac{9}{5}$, 1390 as the equivalent of the thermal unit Centigrade, which is taken as the value of J in the numerical applications contained in the present paper. [Note of Jan. 12, 1882. Joule's recent redetermination gives 771·8 Manchester foot-pounds as the work required to warm 1 lb. of water from 32° to 33° Fahr.]

30. With regard to the determination of the values of μ for different temperatures, it is to be remarked that equation (4) shows that this might be done by experiments upon any substance whatever of indestructible texture, and indicates exactly the experimental data required in each case. For instance, by first supposing the medium to be air; and again, by supposing it to consist partly of liquid water and partly of saturated vapour, we deduce, as is shown in Part III. of this paper, the two expressions (6), given in § 30 of my former paper ("Account of Carnot's Theory"), for the value of μ at any temperature. As yet no experiments have been made upon air which afford the required data for calculating the value of μ through any extensive range of temperature; but for temperatures between 50° and 60° Fahr., Joule's experiments* on the heat evolved by the expenditure of a given amount of work on the compression of air kept at a constant temperature, afford the most direct data for this object which have yet been obtained; since, if Q be the quantity of heat evolved by the compression of a fluid subject to "the gaseous laws" of expansion and compressibility, W the amount of mechanical work spent, and t the constant temperature of the fluid, we have by (11) of § 49 of my former paper,

$$\mu = \frac{W.E}{Q(1+Et)} \quad \text{.......................(10)},$$

* "On the Changes of Temperature produced by the Rarefaction and Condensation of Air," *Phil. Mag.*, Vol. XXVI., May, 1845.

which is in reality a simple consequence of the other expression for μ in terms of data with reference to air. Remarks upon the determination of μ by such experiments, and by another class of experiments on air originated by Joule, are reserved for a separate communication, which I hope to be able to make to the Royal Society on another occasion. [*Dyn. Theory of Heat,* below, Part IV. §§ 61—80.]

31. The second of the expressions (6), in § 30 of my former paper, or the equivalent expression (32), given below in the present paper, shows that μ may be determined for any temperature from determinations for that temperature of—

(1) The rate of variation with the temperature, of the pressure of saturated steam.

(2) The latent heat of a given weight of saturated steam.

(3) The volume of a given weight of saturated steam.

(4) The volume of a given weight of water.

The last mentioned of these elements may, on account of the manner in which it enters the formula, be taken as constant, without producing any appreciable effect on the probable accuracy of the result.

32. Regnault's observations have supplied the first of the data with very great accuracy for all temperatures between -32^0 Cent. and 230^0.

33. As regards the second of the data, it must be remarked that all experimenters, from Watt, who first made experiments on the subject, to Regnault, whose determinations are the most accurate and extensive that have yet been made, appear to have either explicitly or tacitly assumed the same principle as that of Carnot which is overturned by the dynamical theory of heat; inasmuch as they have defined the "total heat of steam" as the quantity of heat required, to convert a unit of weight of water at 0^0, into steam in the particular state considered. Thus Regnault, setting out with this definition for "the total heat of saturated steam," gives experimental determinations of it for the entire range of temperatures from 0^0 to 230^0; and he deduces the

"latent heat of saturated steam" at any temperature, from the "total heat," so determined, by subtracting from it the quantity of heat necessary to raise the liquid to that temperature. Now, according to the dynamical theory, the quantity of heat expressed by the preceding definition depends on the manner (which may be infinitely varied) in which the specified change of state is effected; differing in different cases by the thermal equivalents of the differences of the external mechanical effect produced in the expansion. For instance, the total quantity of heat required to evaporate a quantity of water at 0°, and then, keeping it always in the state of saturated vapour *, bring it to the temperature 100°, cannot be so much as three-fourths of the quantity required, first, to raise the temperature of the liquid to 100°, and then evaporate it at that temperature; and yet either quantity is expressed by what is generally received as a *definition* of the "total heat" of the saturated vapour. To find what it is that is really determined as "total heat" of saturated steam in Regnault's researches, it is only necessary to remark, that the measurement actually made is of the quantity of heat emitted by a certain weight of water in passing through a calorimetrical apparatus, which it enters as saturated steam, and leaves in the liquid state, the result being reduced to what would have been found if the final temperature of the water had been exactly 0°. For there being no external mechanical effect produced (other than that of sound, which it is to be presumed is quite inappreciable), the only external effect is the emission of heat. This must, therefore, according to the fundamental proposition of the dynamical theory, be independent of the intermediate agencies. It follows that, however the steam may rush through the calorimeter, and at whatever reduced pressure it may actually be condensed †, the

* See below (Part III. § 58), where the "negative" specific heat of saturated steam is investigated. If the mean value of this quantity between 0° and 100° were − 1·5 (and it cannot differ much from this) there would be 150 units of heat emitted by a pound of saturated vapour in having its temperature raised (by compression) from 0° to 100°. The latent heat of the vapour at 0° being 606·5, the final quantity of heat required to convert a pound of water at 0° into saturated steam at 100°, in the first of the ways mentioned in the text, would consequently be 456·5, which is only about $\frac{5}{7}$ of the quantity 637 found as "the total heat" of the saturated vapour at 100°, by Regnault.

† If the steam have to rush through a long fine tube, or through a small aperture within the calorimetrical apparatus, its pressure will be diminished before it is

heat emitted externally must be exactly the same as if the condensation took place under the full pressure of the entering saturated steam; and we conclude that *the total heat*, as actually determined from his experiments by Regnault, is the quantity of heat that would be required, first to raise the liquid to the specified temperature, and then to evaporate it at that temperature; and that the principle on which he determines the latent heat is correct. Hence, through the range of his experiments, that is from 0^0 to 230^0, we may consider the second of the data required for the calculation of μ as being supplied in a complete and satisfactory manner.

34. There remains only the third of the data, or the volume of a given weight of saturated steam, for which accurate experiments through an extensive range are wanting; and no experimental researches bearing on the subject having been made since the time when my former paper was written, I see no reason for supposing that the values of μ which I then gave are not the most probable that can be obtained in the present state of science; and, on the understanding stated in § 33 of that paper, that accurate experimental determinations of the densities of saturated steam at different temperatures may indicate considerable errors in the densities which have been assumed according to the "gaseous laws," and may consequently render considerable alterations in my results necessary, I shall still continue to use Table I.

condensed; and there will, therefore, in two parts of the calorimeter be saturated steam at different temperatures (as, for instance, would be the case if steam from a high pressure boiler were distilled into the open air); yet, on account of the heat developed by the fluid friction, which would be precisely the equivalent of the mechanical effect of the expansion wasted in the rushing, the heat measured by the calorimeter would be precisely the same as if the condensation took place at a pressure not appreciably lower than that of the entering steam. The circumstances of such a case have been overlooked by Clausius (Poggendorff's *Annalen*, 1850, No. 4, p. 510), when he expresses with some doubt the opinion that the latent heat of saturated steam will be truly found from Regnault's "total heat," by deducting "the sensible heat;" and gives as a reason that, in the actual experiments, the condensation must have taken place "under the same pressure, or nearly under the same pressure," as the evaporation. The question is not, *Did the condensation take place at a lower pressure than that of the entering steam?* but, *Did* Regnault *make the steam work an engine in passing through the calorimeter, or was there so much noise of steam rushing through it as to convert an appreciable portion of the total heat into external mechanical effect?* And a negative answer to this is a sufficient reason for adopting *with certainty* the opinion that the principle of his determination of the latent heat is correct.

of that paper, which shows the values of μ for the temperatures $\frac{1}{2}$, $1\frac{1}{2}$, $2\frac{1}{2}$...$230\frac{1}{2}$, or, the mean values of μ for each of the 230 successive Centrigrade degrees of the air-thermometer above the freezing-point, as the basis of numerical applications of the theory. It may be added, that any experimental researches sufficiently trustworthy in point of accuracy, yet to be made, either on air or any other substance, which may lead to values of μ differing from those, must be admitted as proving a discrepancy between the true densities of saturated steam, and those which have been assumed*.

35. Table II. of my former paper, which shows the values of $\int_0^t \mu dt$ for $t=1$, $t=2$, $t=3$, ... $t=231$, renders the calculation of the mechanical effect derivable from a given quantity of heat by means of a perfect engine, with any given range included between the limits 0 and 231, extremely easy; since the quantity to be divided by J† in the index of the exponential in the expression (8) will be found by subtracting the number in that table corresponding to the value of T, from that corresponding to the value of S.

36. The following tables show some numerical results which have been obtained in this way, with a few (contained in the lower part of the second table) calculated from values of $\int_0^t \mu dt$

* I cannot see that any hypothesis, such as that adopted by Clausius fundamentally in his investigations on this subject, and leading, as he shows to determinations of the densities of saturated steam at different temperatures, which indicate enormous deviations from the gaseous laws of variation with temperature and pressure, is more probable, or is probably nearer the truth, than that the density of saturated steam does follow these laws as it is usually assumed to do. In the present state of science it would perhaps be wrong to say that either hypothesis is more probable than the other [or that the rigorous truth of either hypothesis is probable at all].

† It ought to be remarked, that as the unit of force implied in the determinations of μ is the weight of a pound of matter at Paris, and the unit of force in terms of which J is expressed is the weight of a pound at Manchester, these numbers ought in strictness to be modified so as to express the values in terms of a common unit of force; but as the force of gravity at Paris differs by less than $\frac{1}{2000}$ of its own value from the force of gravity at Manchester, this correction will be much less than the probable errors from other sources, and may therefore be neglected.

estimated for temperatures above 230°, roughly, according to the rate of variation of that function within the experimental limits.

37. *Explanation of the Tables.*

Column I. in each table shows the assumed ranges.

Column II. shows ranges deduced by means of Table II. of the former paper, so that the value of $\int_T^S \mu dt$ for each may be the same as for the corresponding range shown in column I.

Column III. shows what would be the duty of a unit of heat if Carnot's theory required no modification (or the actual duty of a unit of heat with additions through the range, to compensate for the quantities converted into mechanical effect).

Column IV. shows the true duty of a unit of heat, and a comparison of the numbers in it with the corresponding numbers in column III. shows how much the true duty falls short of Carnot's theoretical duty in each case.

Column VI. is calculated by the formula

$$R = \epsilon^{-\frac{1}{1390}\int_T^S \mu dt},$$

where $\epsilon = 2{\cdot}71828$, and for $\int_T^S \mu dt$ the successive values shown in column III. are used.

Column IV. is calculated by the formula

$$W = 1390\,(1 - R)$$

from the values of $1 - R$ shown in column V.

38. *Table of the Motive Power of Heat.*

Range of temperatures.				III.	IV.	V.	VI.
I.		II.		Duty of a unit of heat through the whole range.	Duty of a unit of heat supplied from the source.	Quantity of heat converted into mechanical effect.	Quantity of heat wasted.
S.	*T.*	*S.*	*T.*	$\int_T^S \mu dt.$	*W.*	1—*R.*	*R.*
°		°	°	ft.-lbs	ft.-lbs.		
1	0	31·08	30	4·960	4·948	·00356	·99644
10	0	40·86	30	48·987	48·1	·0346	·9654
20	0	51·7	30	96·656	93·4	·067	·933
30	0	62·6	30	143·06	136	·098	·902
40	0	73·6	30	188·22	176	·127	·873
50	0	84·5	30	232·18	214	·154	·846
60	0	95·4	30	274·97	249	·179	·821
70	0	106·3	30	316·64	283	·204	·796
80	0	117·2	30	357·27	315	·227	·773
90	0	128·0	30	396·93	345	·248	·752
100	0	138·8	30	435·69	374	·269	·731
110	0	149·1	30	473·62	401	·289	·711
120	0	160·3	30	510·77	427	·308	·692
130	0	171·0	30	547·21	452	·325	·675
140	0	181·7	30	582·98	476	·343	·657
150	0	192·3	30	618·14	499	·359	·641
160	0	203·0	30	652·74	521	·375	·625
170	0	213·6	30	686·80	542	·390	·610
180	0	224·2	30	720·39	562	·404	·596
190	0	190	0	753·50	582	·418	·582
200	0	200	0	786·17	600	·432	·568
210	0	210	0	818·45	619	·445	·555
220	0	220	0	850·34	636	·457	·542
230	0	230	0	881 87	653	·470	·530

39. *Supplementary Table of the Motive Powers of Heat.*

Range of temperatures.				III.	IV.	V.	VI.
I.		II.		Duty of a unit of heat through the whole range.	Duty of a unit of heat supplied from the source.	Quantity of heat converted into mechanical effect.	Quantity of heat wasted.
S.	*T.*	*S.*	*T.*	$\int_T^S \mu dt.$	*W.*	1—*R.*	*R.*
°	°	°	°	ft.-lbs.	ft.-lbs.		
101·1	0	140	30	439·9	377	·271	·729
105·8	0	230	100	446·2	382	·275	·725
300	0	300	0	1099	757	·545	·455
400	0	400	0	1395	879	·632	·368
500	0	500	0	1690	979	·704	·296
600	0	600	0	1980	1059	·762	·238
∞	0	∞	0	∞	1390	1·000	·000

40. Taking the range 30° to 140° as an example suitable to the circumstances of some of the best steam-engines that have yet

been made (see Appendix to "Account of Carnot's Theory," Sec. v.), we find in column III. of the supplementary table, 377 ft.-lbs. as the corresponding duty of a unit of heat instead of 440, shown in column III., which is Carnot's theoretical duty. We conclude that the recorded performance of the Fowey-Consols engine in 1845, instead of being only $57\frac{1}{2}$ per cent. amounted really to 67 per cent., or $\frac{2}{3}$ of the duty of a perfect engine with the same range of temperature; and this duty being ·271 (rather more than $\frac{1}{4}$) of the whole equivalent of the heat used; we conclude further, that $\frac{1}{5{\cdot}49}$, or 18 per cent. of the whole heat supplied, was actually converted into mechanical effect by that steam-engine.

41. The numbers in the lower part of the supplementary table show the great advantage that may be anticipated from the perfecting of the air-engine, or any other kind of thermo-dynamic engine in which the range of the temperature can be increased much beyond the limits actually attainable in steam-engines. Thus an air-engine, with its hot part at 600°, and its cold part at 0° Cent., working with perfect economy, would convert 76 per cent. of the whole heat used into mechanical effect; or working with such economy as has been estimated for the Fowey-Consols engine, that is, producing 67 per cent. of the theoretical duty corresponding to its range of temperature, would convert 51 per cent. of all the heat used into mechanical effect. [Note, of Dec. 30, 1881. A great advance towards realizing this principle is now achieved in the gas-engine, of which the true dynamical economy is believed to be already superior to that of the best modern compound steam-engine.]

42. It was suggested to me by Mr Joule, in a letter dated December 9, 1848, that the true value of μ might be "inversely as the temperatures from zero *;" and values for various temperatures calculated by means of the formula,

$$\mu = J\frac{E}{1+Et} \text{.........................(11)},$$

* If we take $\mu = k\frac{E}{1+Et}$ where k may be any constant, we find

$$W = J\left(\frac{S-T}{\frac{1}{E}+S}\right)^{\frac{k}{J}};$$

which is the formula I gave when this paper was communicated. I have since remarked, that Mr Joule's hypothesis implies essentially that the coefficient k must were given for comparison with those which I had calculated from data regarding steam. This formula is also adopted by Clausius, who uses it fundamentally in his mathematical investigations. If μ were correctly expressed by it, we should have

$$\int_T^S \mu dt = J \log \frac{1+ES}{1+ET};$$

and therefore equations (1) and (2) would become

$$W = J\frac{S-T}{\frac{1}{E}+S} \qquad (12),$$

$$R = \frac{\frac{1}{E}+T}{\frac{1}{E}+S} \qquad (13).$$

43. The reasons upon which Mr Joule's opinion is founded, that the preceding equation (11) may be the correct expression for Carnot's function, although the values calculated by means of it differ considerably from those shown in Table I. of my former paper, form the subject of a communication which I hope to have an opportunity of laying before the Royal Society previously to the close of the present session.

[*Editor's Note:* Material has been omitted at this point.]

be as it is taken in the text, the mechanical equivalent of a thermal unit. Mr Rankine, in a letter dated March 27, 1851, informs me that he has deduced, from the principles laid down in his paper communicated last year to this Society, an approximate formula for the ratio of the maximum quantity of heat converted into mechanical effect to the whole quantity expended, in an expansive engine of any substance, which, on comparison, I find agrees exactly with the expression (12) given in the text as a consequence of the hypothesis suggested by Mr Joule regarding the value of μ at any temperature.—[April 4, 1851.]

7

Reprinted from *Philos. Mag.*, ser. 4, **24**(159, 160) 81–97, 201–213(1862)

ON THE APPLICATION OF THE THEOREM OF THE EQUIVALENCE OF TRANSFORMATIONS TO THE INTERNAL WORK OF A MASS OF MATTER

R. Clausius*

IN a memoir published in the year 1854†, wherein I sought to simplify to some extent the form of the developments I had previously published, I deduced, from my fundamental proposition *that heat cannot of itself pass from a colder into a warmer body*, a principle which is closely allied to, but does not entirely coincide with, the one first deduced by S. Carnot from considerations of a different class, based upon the older views of the nature of heat. It has reference to the circumstances under which work can be transformed into heat, and, conversely, heat converted into work; and I have called it the *Principle of the Equivalence of Transformations*. I did not, however, there communicate the entire proposition in the general form in which I had deduced it, but confined myself on that occasion to the publication of a part which can be treated separately from the rest, and is capable of more strict proof.

In general, when a body changes its state, work is performed *externally* and *internally* at the same time,—the external work having reference to the forces which extraneous bodies exert upon the body under consideration, and the internal work to the forces exerted by the constituent molecules of the body in question upon each other. The internal work is for the most part so little known, and connected with another equally unknown quan-

* Translated from the *Mittheilungen der Naturforschenden Gesellschaft in Zürich*, vol. vii. p. 48, having been communicated to the Society on the 27th of January, 1862.

† "On a modified form of the second Fundamental Theorem in the Mechanical Theory of Heat" (Phil. Mag. S. 4. vol. xii. p. 81; Pogg. *Ann*. vol. xciii. p. 481).

tity in such a way, that in treating of it we are obliged in some measure to trust to probabilities; whereas the external work is immediately accessible to observation and measurement, and thus admits of more strict treatment. Accordingly, since, in my former paper, I wished to avoid everything that was hypothetical, I entirely excluded the internal work, which I was able to do by confining myself to the consideration of *circular processes* —that is to say, operations in which the modifications which the body undergoes are so arranged that the body finally returns to its original condition. In such operations the internal work which is performed during the separate modifications, partly in a positive sense and partly in a negative sense, neutralizes itself, so that nothing but external work remains, for which the principle in question can then be demonstrated with mathematical strictness, starting from the above-mentioned fundamental proposition.

I have delayed till now the publication of the remainder of my theorem, because it leads to a consequence which is considerably at variance with the ideas hitherto generally entertained of the heat contained in bodies, and I therefore thought it desirable to make still further trial of it. But as I have become more and more convinced in the course of years that we must not attach too great weight to such ideas, which in part are founded more upon usage than upon a scientific basis, I feel that I ought to hesitate no longer, but to submit to the scientific public the theorem of the equivalence of transformations in its complete form, with the principles which attach themselves to it. I venture to hope that the importance which these principles, supposing them to be true, possess in connexion with the theory of heat will be thought to justify their publication in their present hypothetical form.

I will, however, at once distinctly observe that, whatever hesitation may be felt in admitting the truth of the following principles, the conclusions arrived at in my former paper, in reference to circular processes, lose thereby none of their authority.

§ 1. I will begin by briefly stating the principle of the equivalence of transformations, as I have already developed it, in order to be able to connect with it the following considerations.

When a body goes through a circular process, a certain amount of external work may be gained, in which case a certain quantity of heat must be simultaneously expended; or, conversely, work may be expended and a corresponding quantity of heat may be gained. This may be expressed by saying:—*Heat can be transformed into work, or work into heat, by a circular process.*

There may also be another effect of a circular process: heat may be transferred from one body to another, by the body which is undergoing modification absorbing heat from the one body and giving it out again to the other. In this case the bodies between which the transference of heat takes place are to be viewed merely as heat-reservoirs, of which we are not concerned to know anything except the temperatures. If the temperatures of the two bodies differ, heat passes, either from a warmer to a colder body, or from a colder to a warmer body, according to the direction in which the transference of heat takes place. Such a passage of heat may also be designated, for the sake of uniformity, as a *transformation*, inasmuch as it may be said that *heat of one temperature is transformed into heat of another temperature.*

The two kinds of transformations that have been mentioned are related in such a way that one presupposes the other, and that they can mutually replace each other. If we call transformations which can replace each other *equivalent*, and seek the mathematical expressions which determine the amount of the transformations in such a manner that equivalent transformations become equal in magnitude, we arrive at the following expression :—*If the quantity of heat* Q *of the temperature* t *is produced from work, the equivalent value of this transformation is*

$$\frac{Q}{T};$$

and if the quantity of heat Q *passes from a body whose temperature is* t_1 *into another whose temperature is* t_2, *the equivalent value of this transformation is*

$$Q\left(\frac{1}{T_2}-\frac{1}{T_1}\right),$$

where T is a function of the temperature which is independent of the kind of process by means of which the transformation is effected, and T_1 and T_2 denote the values of this function which correspond to the temperatures t_1 and t_2. I have shown by separate considerations that T is in all probability nothing more than the absolute temperature.

These two expressions further enable us to recognize the positive or negative sense of the transformations. In the first, Q is taken as positive when work is transformed into heat, and as negative when heat is transformed into work. In the second, we may always take Q as positive, since the opposite senses of the transformations are indicated by the possibility of the difference $\frac{1}{T_2}-\frac{1}{T_1}$ being either positive or negative. It will thus be seen that the passage of heat from a higher to a lower temperature is to be looked upon as a positive transformation, and its passage from a lower to a higher temperature as a negative transformation.

If we represent the transformations which occur in a circular process by these expressions, the relation existing between them can be stated in a simple and definite manner. If the circular process is *reversible*, the transformations which occur therein must be partly positive and partly negative, and the equivalent values of the positive transformations must be together equal to those of the negative transformations, so that the algebraic sum of all the equivalent values becomes $=0$. If the circular process is *not reversible*, the equivalent values of the positive and negative transformations are not necessarily equal, but they can only differ in such a way that the positive transformations predominate. The proposition respecting the equivalent values of the transformations may accordingly be stated thus:—*The algebraic sum of all the transformations occurring in a circular process can only be positive, or, as an extreme case, equal to nothing.*

The mathematical expression for this proposition is as follows. Let dQ be an element of the heat given up by the body to any reservoir of heat during its modifications (heat which it may absorb from a reservoir being here reckoned as negative), and T the absolute temperature of the body at the moment of giving up this heat, then the equation

$$\int \frac{dQ}{T} = 0 \quad . \quad . \quad . \quad . \quad . \quad . \quad . \quad . \quad (\text{I.})$$

must be true for every reversible circular process, and the relation

$$\int \frac{dQ}{T} \geqq 0 \quad . \quad . \quad . \quad . \quad . \quad . \quad . \quad . \quad (\text{I}a.)$$

must hold good for every circular process which is in any way possible.

§ 2. Although the necessity of this proposition admits of strict mathematical proof if we start from the fundamental principle above quoted, it thereby nevertheless retains an abstract form, in which it is difficultly embraced by the mind, and we feel compelled to seek for the precise physical cause, of which this proposition is a consequence. Moreover, since there is no essential difference between internal and external work, we may assume almost with certainty that a proposition which is so generally applicable to external work cannot be restricted to this alone, but that, where external work is combined with internal work, it must be capable of application to the latter also.

Considerations of this nature led me, in my first investigations into the mechanical theory of heat, to assume a general law respecting the dependence of the active force of heat on temperature, among the immediate consequences of which is the principle of the equivalence of transformations in its more complete form, and which at the same time involves other important conclusions. This law I will at once quote, and will endeavour to make its meaning clear by the addition of a few comments. As for the reasons for supposing it to be true, such as do not at once appear from its internal probability will gradually become apparent in the course of this paper. It is as follows:—

In all cases in which the heat contained in a body does mechanical work by overcoming a resistance, the magnitude of the resistance which it is capable of overcoming is proportional to the absolute temperature.

In order to understand the significance of this law, we require to consider more closely the processes by which heat can perform mechanical work. These processes always admit of being reduced to the alteration in some way or another of the arrangement of the constituent molecules of a body. For instance, bodies are expanded by heat, their molecules being thus separated from each other: in this case the mutual attractions of the molecules on the one hand, and on the other external opposing forces, in so far as any such are in operation, have to be overcome. Again, the state of aggregation of bodies is altered by heat, solid bodies being rendered liquid, and both solid and liquid bodies being rendered aëriform: here likewise internal forces, and in general external forces also, have to be overcome. Another case which I will also mention, because it differs so widely from the foregoing, and therefore shows how various are the modes of action which belong to the class we are considering, is the transference of electricity from one body to the other, constituting the thermoelectric current, which takes place by the action of heat on two heterogeneous bodies in contact.

In the cases first mentioned, the arrangement of the molecules is altered. Since, even while a body remains in the same state of aggregation, its molecules do not retain fixed unvarying positions, but are constantly in a state of more or less extended motion, we may, when speaking of the *arrangement of the molecules* at any particular time, understand either the arrangement which would result from the molecules being fixed in the actual positions they occupy at the instant in question, or we may suppose such an arrangement that each molecule occupies its mean position. Now the effect of heat always tends to loosen the connexion between the molecules, and so to increase their mean distances from one another. In order to be able to represent this mathematically, we will express the degree in which the molecules of a body are dispersed, by introducing a new magnitude, which we will call the *disgregation* of the body, and by help of which we can define the effect of heat as simply *tending to increase the disgregation.* The way in which a definite measure of this magnitude can be arrived at will appear from the sequel.

In the case last mentioned, an alteration in the arrangement of the electricity takes place, an alteration which can be represented and taken into calculation in a way corresponding to the alteration of the position of the molecules, and which, when it occurs, we will consider as always included in the general expression *alteration of arrangement,* or *alteration of disgregation.*

It is evident that each of the kinds of alteration that have been named may also take place in the reverse sense, if the effect of the opposing forces is greater than that of the heat. We will assume as likewise self-evident that, for the production of work, a corresponding quantity of heat must always be expended, and conversely, that, by the consumption of work, an equivalent quantity of heat must be produced.

§ 3. If we now consider more closely the various cases which occur in relation to the forces which are operative in each of them, the case of the expansion of a permanent gas presents itself as particularly simple. We may conclude from certain properties of the gases that the mutual attraction of their molecules at their mean distances is very small, and therefore that only a very slight resistance is offered to the expansion of a gas, so that the resistance of the sides of the containing vessel must maintain equilibrium with almost the whole effect of the heat. Accordingly the externally sensible pressure of a gas forms an approximate measure of the separative force of the heat contained in the gas; and hence, according to the foregoing proposition, this pressure must be nearly proportional to the absolute temperature. The internal probability of the truth of this result is indeed so great, that many physicists since Gay-Lussac and Dalton have without hesitation presupposed this proportionality, and have employed it for calculating the absolute temperature.

In the above-mentioned case of thermo-electric action, the force which exerts an action contrary to that of the heat is likewise simple and easily determined. For at the point of contact of two heterogeneous substances, such a quantity of electricity is driven from the one to the other by the action of the heat, that the opposing force resulting from the electric tension suffices to hold the force exerted by the heat in equilibrium. Now in a former memoir "On the application of the Mechanical Theory of Heat to the Phenomena of Thermal Electricity*," I have shown that, in so far as changes in the arrangement of the molecules are not necessarily produced at the same time by changes of temperature, the difference of tension produced by heat must be proportional

* Poggendorff's *Annalen,* vol. xc. p. 513.

to the absolute temperature, as is required by the foregoing theorem.

In the other cases that are quoted, as well as in most others, the relations are less simple, because in them an essential part is played by the forces exerted by the molecules upon one another, forces which, as yet, are quite unknown. It results, however, from the mere consideration of the external resistances which heat is capable of overcoming, that in general its force increases with the temperature. If we wish, for instance, to prevent the expansion of a body by means of external pressure, we are obliged to employ a greater pressure the more the body is heated; hence we may conclude, without having a knowledge of the internal forces, that the total amount of the resistances which can be overcome in expansion, increases with the temperature. We cannot, however, directly ascertain whether it increases exactly in the proportion required by the foregoing theorem, without knowing the internal forces. On the other hand, if this theorem be regarded as proved on other grounds, we may reverse the process, and employ it for the determination of the internal forces exerted by the molecules.

The forces exerted upon one another by the molecules are not of so simple a kind that each molecule can be replaced by a mere point; for many cases occur in which it can be easily seen that we have not merely to consider the distances of the molecules, but also their relative positions. If we take, for example, the melting of ice, there is no doubt that here internal forces, exerted by the molecules upon each other, are overcome, and accordingly disgregation takes place; nevertheless the centres of gravity of the molecules are on the average not so far removed from each other in the liquid water as they were in the ice, for the water is the more dense of the two. Again, the peculiar behaviour of water in contracting when heated above 0° C., and only beginning to expand when its temperature exceeds 4°, shows that likewise in liquid water, in the neighbourhood of its melting-point, increase of disgregation is not connected with increase of the mean distances of its molecules. In the case of the internal forces, it would accordingly be difficult—even if we did not want to measure them, but only to represent them mathematically—to find a fitting expression for them which would admit of a simple determination of magnitude. This difficulty, however, disappears if we take into calculation, not the forces themselves, but the *mechanical work* which in any alteration of arrangement is required to overcome them. The expressions for the quantities of work are simpler than those for the corresponding forces; for the quantities of work can be all expressed, without further secondary statements, by numbers with the same unit, which can be added together, or subtracted from one another, however various the forces may be to which they are referable.

It is therefore convenient to alter the form of the above theorem by introducing, instead of the forces themselves, the work done in overcoming them. In this form it reads as follows:—

The mechanical work which can be exerted by heat in any alteration of the arrangement of a body is proportional to the absolute temperature at which this alteration occurs.

§ 4. The theorem does not speak of the work which the heat *does*, but of the work which it *can do*; and similarly, in the first form of the theorem, the resistances which the heat *can overcome* are spoken of. This distinction is necessary for the following reasons.

Since the external forces which act upon a body while it is undergoing an alteration of arrangement may vary very greatly, it may happen that the heat, while causing an alteration of arrangement, has not to overcome the whole resistance which it would be possible for it to overcome. A well-known and often-quoted example of this is afforded by a gas which expands under such conditions that it has not to overcome an opposing pressure equal to its own expansive force, as, for instance, when the space filled by the gas is made to communicate with another which is empty, or contains a gas of lower pressure. In order in such cases to determine the force of the heat, we must evidently not consider the resistance which actually is overcome, but that which can be overcome.

Also in alterations of arrangement of the opposite kind, that is, where the action of heat is overcome by the opposing forces, a similar distinction may require to be made, but in this case only as far as this—that the total amount of the forces by which the action of the heat is overcome may be greater than the active force of the heat, but not smaller.

Cases in which these differences occur may be thus characterized. When an alteration of arrangement takes place so that the force and counterforce are equal, the alteration can take place in the reverse direction also under the influence of the same forces. But if it occurs so that the overcoming force is greater than that which is overcome, the transformation cannot take place in the opposite direction under the influence of the same forces. We may say that the transformation has occurred in the first case in a *reversible* manner, and in the second case in an *irreversible* manner.

Strictly speaking, the overcoming force must always be more powerful than the force which it overcomes; but as the excess of force does not require to have any assignable value, we may think of it as becoming continually smaller and smaller, so that its value may approach to nought as nearly as we please. Hence it may be seen that the case in which the transformations take place reversibly is a limit which in reality is never quite reached, but to which we can approach as nearly as we please. We may therefore, in theoretical discussions, still speak of this case as one which really exists; indeed, as a limiting case it possesses special theoretical importance.

I will take this opportunity of mentioning another process in which likewise this distinction is to be observed. In order for one body to impart heat to another by conduction or radiation (in the case of radiation, wherein mutual communication of heat takes place, it is to be understood that we speak here of a body which gives out more heat than it receives), the body which parts with heat must be warmer than the body which takes up heat; and hence the passage of heat between two bodies of different temperature can take place in one direction only, and not in the contrary direction. The only case in which the passage of heat can occur equally in both directions is when it takes place between bodies of equal temperature. Strictly speaking, however, the communication of heat from one body to another of the same temperature is not possible; but since the difference of temperature may be as small as we please, the case in which it is equal to nothing, and the passage of heat accordingly reversible, is a limiting case which may be regarded as theoretically possible.

§ 5. We will now deduce the mathematical expression for the above theorem, treating in the first place the case in which the change of condition undergone by the body under consideration takes place *reversibly*. The result at which we shall arrive for this case will easily admit of subsequent generalization, so as to include also the cases in which an alteration occurs irreversibly.

Let the body be supposed to undergo an infinitely small alteration of condition, whereby the quantity of heat contained in it, and also the arrangement of its constituent molecules, may be altered. Let the quantity of heat contained in it be expressed by H, and the alteration of this quantity by dH. Further, let the work, both internal and external together, performed by the heat in the change of arrangement be denoted by dL, a magnitude which may be either positive or negative according as the active force of the heat overcomes the forces acting in the contrary direction, or is overcome by them. We obtain the heat expended to produce this quantity of work by multiplying the work by the heat-equivalent of a unit of work which we may call A; hence it is AdL.

The sum $dH + AdL$ is the quantity of heat which the body must receive from without, and must accordingly withdraw from another body during the alteration of condition. We have, however, already represented by dQ the infinitely small quantity of heat imparted to another body by the one which is undergoing modification, hence we must represent in a corresponding manner, by $-dQ$, the heat which it withdraws from another body. We thus obtain the equation

$$-dQ = dH + AdL,$$

or

$$dQ + dH + AdL = 0^*. \quad . \quad . \quad . \quad . \quad (1)$$

In order now to be able to introduce the disgregation also into the formulæ, we must first settle how we are to determine it as a mathematical quantity.

* In my previous memoirs I have separated from one another the *internal* and the *external* work performed by the heat during the change of condition of the body. If the former be denoted by dI, and the latter by dW, the above equation becomes

$$dQ + dH + AdI + AdW = 0. \quad . \quad . \quad . \quad . \quad . \quad . \quad . \quad (a)$$

Since, however, the increase in the quantity of heat actually contained in a body, and the heat consumed by internal work during an alteration of condition, are magnitudes of which we commonly do not know the individual values, but only the sum of those values, and which resemble each other in being fully determined as soon as we know the initial and final conditions of the body, without our requiring to know how it has passed from the one to the other, I have thought it advisable to introduce a function which shall represent the sum of these two magnitudes, and which I have denoted by U. Accordingly

$$dU = dH + AdI, \quad . \quad . \quad . \quad . \quad . \quad . \quad . \quad . \quad . \quad . \quad (b)$$

and hence the foregoing equation becomes

$$dQ + dU + AdW = 0; \quad . \quad . \quad . \quad . \quad . \quad . \quad . \quad (c)$$

and if we suppose the last equation integrated for any finite alteration of condition, we have

$$Q + U + AW = 0. \quad . \quad . \quad . \quad . \quad . \quad . \quad . \quad . \quad . \quad (d)$$

These are the equations which I have used in my memoirs published in 1850 and in 1854, partly in the particular form which they assume for the permanent gases, and partly in the general form in which they are here given, with no other difference than that I there took the positive and negative quantities of heat in the opposite sense to what I have done here, in order to attain greater correspondence with the equation (I.) given in § 1. The function U which I introduced is capable of manifold application in the theory of heat, and, since its introduction, has been the subject of very interesting mathematical developments by W. Thomson and by Kirchhoff (see Philosophical Magazine, S. 4. vol. ix. p. 523, and Poggendorff's *Annalen*, vol. ciii. p. 177). Thomson has called it "the mechanical energy of a body in a given state," and Kirchhoff "Wirkungsfunction." Although I consider my original definition of it (see Pogg. *Ann.* vol. lxxix. p. 385, and vol. xciii. p. 484), as representing the *sum of the heat added to the quantity already present and of that expended in internal work*, starting from any given initial state, as perfectly exact, I can still have no objection to make against an abbreviated mode of expression.

By disgregation is represented, as stated in § 2, the degree of dispersion of the body. Thus, for example, the disgregation of a body is greater in the liquid state than in the solid, and greater in the aëriform than in the liquid state. Further, if part of a given quantity of matter is solid and the rest liquid, the disgregation is greater the greater the proportion of the whole mass that is liquid; and similarly, if one part is liquid and the remainder aëriform, the disgregation is greater the larger the aëriform portion. The disgregation of a body is fully determined when the arrangement of its constituent molecules is given; but, on the other hand, we cannot say conversely that the arrangement of the constituent molecules is determined when the magnitude of the disgregation is known. It might, for example, happen that the disgregation of a given quantity of matter should be the same when one part was solid and one part aëriform, as when the whole mass was liquid.

We will now suppose that, with the aid of heat, the body changes its state, and we will provisionally confine ourselves to such changes of state as can occur in a constant and reversible manner, and we will also assume that the body has a uniform temperature throughout. Since the increase of disgregation is the action by means of which heat performs work, it follows that the quantity of work must bear a definite ratio to the quantity by which the disgregation is increased; we will therefore fix the still arbitrary determination of the magnitude of disgregation so that, at any given temperature, the increase of disgregation shall be proportional to the work which the heat could perform at that temperature. All that further regards the influence of temperature is determined by the foregoing theorem. For if the same alteration of disgregation takes place at different temperatures, the corresponding work must be proportional to the absolute temperature. Accordingly, let Z be the disgregation of the body, and dZ an infinitely small alteration of it, and let dL be the corresponding infinitely small quantity of work, we can then put

$$dL = KTdZ,$$

or

$$dZ = \frac{dL}{KT},$$

where K is a constant dependent on the unit, hitherto left undecided upon, according to which Z is to be measured. We will choose this unit of measure so that $K = \frac{1}{A}$, and the equation becomes

$$dZ = \frac{AdL}{T} \qquad (2)$$

If we suppose this expression integrated, starting with any initial condition in which Z has the value Z_0, we get

$$Z = Z_0 + A\int\frac{dL}{T}. \quad . \quad . \quad . \quad . \quad . \quad . \quad (3)$$

The magnitude Z is thus determined, with the exception of a constant dependent upon the initial condition that is chosen.

If the temperature of the body is not the same at every part, we can regard it as divided into any number we choose of separate parts, and let the elements dZ and dL in equation (2) refer to any one of them, and at once substitute for T the value of the absolute temperature of that part. If we then unite by summation the infinitely small changes of disgregation of the separate parts, or by integration, if there is an infinite number of them, we obtain the similarly infinitely small change of disgregation of the entire body, and from this we can obtain, likewise by integration, any desired finite change of disgregation.

We will now return to equation (1), and by help of equation (2) we will eliminate from it the element of work dL. Thus we get

$$dQ + dH + TdZ = 0 ; \quad . \quad . \quad . \quad . \quad . \quad . \quad (4)$$

or, dividing by T,

$$\frac{dQ + dH}{T} + dZ = 0. \quad . \quad . \quad . \quad . \quad . \quad . \quad . \quad (5)$$

If we suppose this equation integrated for a finite change of condition, we have

$$\int\frac{dQ + dH}{T} + \int dZ = 0. \quad . \quad . \quad . \quad . \quad . \quad (II.)$$

Supposing the body not to be of uniform temperature throughout, we may imagine it broken up again into separate parts, and can make the elements dQ, dH, and dZ in equation (5) refer in the first instance to one part, and for T we can put the absolute temperature of this part. The symbols of integration in (II.) are then to be understood as embracing the alterations of all the parts. We must here remark that cases in which one continuous body is of different temperatures at different parts, so that a passage of heat immediately takes place by conduction from the warmer to the colder parts, must be for the present disregarded, because such a passage of heat is not reversible, and we have provisionally confined ourselves to the consideration of reversible alterations.

Equation (II.) is the mathematical expression of the above theorem for which we have been seeking, *for all reversible alterations of condition of a body*; and it is clearly evident that it also remains applicable, if a series of successive alterations of condition be considered instead of a single one.

§ 6. The differential equation (4), whence equation (II.) is derived, is connected with a differential equation which results from the already known principles of the mechanical theory of heat, and which transforms itself directly into (4) for the particular case in which the body under consideration is a perfect gas.

We will suppose that there is given any body of variable volume, acted upon, as by an external force, by the pressure exerted on the surface. Let the volume which it assumes under this pressure, p, at the temperature T (reckoned from the absolute zero) be v, and let it be supposed that the condition of the body is fully determined by the magnitudes T and v. If we now denote by $\frac{dQ}{dv}dv$ the quantity of heat which the body must take up in order to expand to the extent of dv, without alteration of temperature (for the sake of conformity with the mode in which the signs are used in the other equations occurring in this section, the positive sense of the quantity of heat is here taken differently from what it is in equation (4), in which heat given up by the body, and not heat communicated to it, is reckoned positive), the following well-known equation, from the mechanical theory of heat, will hold good:—

$$\frac{dQ}{dv}=AT\frac{dp}{dT}.$$

Let us now suppose that the temperature of the body is changed by dT, and its volume by dv, and let us call the quantity of heat which it then takes up dQ; we may then write

$$dQ=\frac{dQ}{dT}dT+\frac{dQ}{dv}dv.$$

For the magnitude here denoted by $\frac{dQ}{dT}$, which represents the specific heat with constant volume, we can put the letter c, and for $\frac{dQ}{dv}$ the expression already given. Then we have

$$dQ=c\,dT+AT\frac{dp}{dT}dv. \quad . \quad . \quad . \quad . \quad . \quad . \quad (6)$$

The only external force, which the body has to overcome on expanding, being p, the work which it performs in so doing is pdv, and the magnitude $\frac{dp}{dT}dv$ indicates the increase of this work with the temperature.

If we now apply this equation to a perfect gas, the specific heat under constant volume is in this case to be regarded as the true specific heat, and this gives the increase in the quantity of heat actually present in the gas; for here no heat is consumed in work, since external work is only performed when increase of volume occurs, and internal work has no existence in the case of perfect gases. We may therefore regard $cd\mathrm{T}$ as identical with $d\mathrm{H}$. We have further, for the perfect gases, the equation

$$pv = \mathrm{RT},$$

where R is a constant, and thence we get

$$\frac{dp}{d\mathrm{T}}dv = \frac{\mathrm{R}}{v}dv = \mathrm{R}d \,.\, \log v.$$

Equation (6) is thus transformed into

$$d\mathrm{Q} = d\mathrm{H} + \mathrm{ART}d \,.\, \log v. \quad . \quad . \quad . \quad . \quad . \quad . \quad (7)$$

This equation agrees, disregarding the difference in the sign of $d\mathrm{Q}$ (which is caused only by the different way in which we have chosen to employ the signs + and − in this case), with equation (4), and the function there represented by the general symbol Z has, in this particular case, the form $\mathrm{AR} \log v$.

Rankine, who has written several interesting memoirs on the transformation of heat into work*, has in like manner proposed to transform equation (6), which in its original form applies to perfect gases only†, so as to render it applicable to other bodies, and writes (only with slightly different letters)

$$d\mathrm{Q} = kd\mathrm{T} + \mathrm{AT}d\mathrm{F}, \quad . \quad . \quad . \quad . \quad . \quad . \quad (8)$$

where k denotes the true specific heat of the body, and F is a magnitude in the determination of which Rankine appears to have been led chiefly by the circumstance mentioned above, that the quantity $\frac{dp}{d\mathrm{T}}dv$ which occurs in equation (6) represents the increase of external work which accompanies an infinitely small alteration of state under increased temperature. Rankine defines the magnitude F as "the rate of variation of effective work with temperature;" and denoting the external work which the body can do in passing, at a given temperature, from a given former state into its present condition, by U, he puts

$$\mathrm{F} = \frac{d\mathrm{U}}{d\mathrm{T}}. \quad . \quad . \quad . \quad . \quad . \quad . \quad . \quad . \quad (9)$$

In the discussion which immediately follows, of the case in which

* Philosophical Magazine, S. 4. vol. v. p. 106; Edinburgh New Philosophical Journal, vol. ii. p. 120; Manual of the Steam-engine.

† Manual of the Steam-engine, p. 310.

the external work consists only in overcoming an external pressure, he gives the equation

$$U = \int p dv,$$

whence follows

$$F = \int \frac{dp}{dT} dv. \quad . \quad . \quad . \quad . \quad . \quad . \quad (10)$$

The integrals which here occur are to be taken from a given initial volume to the actually existing volume, the temperature being supposed constant. Introducing this value of F into equation (8), he writes it in the following form :—

$$dQ = \left(k + AT \int_{\infty}^{v} \frac{d^2p}{dT^2} dv\right) dT + AT \frac{dp}{dT} dv. \quad . \quad (11)$$

His reason for taking an infinitely large volume as the initial volume is not stated, although the choice of the initial volume is evidently not indifferent.

It is easy to see that this manner of modifying equation (6) is very different from my development; the results are also discordant; for the quantity F is not identical with the corresponding $\frac{1}{A}Z$ in my equations, but only coincides with it in that part which could be deduced from data already known; that is to say, the last member of equation (6) gives the differential coefficient v for the magnitude which has to be introduced, since, to get the correct value of this member, we must in any case put

$$\frac{dF}{dv} = \frac{1}{A} \frac{dZ}{dv} = \frac{dp}{dT}. \quad . \quad . \quad . \quad . \quad . \quad (12)$$

Rankine has, however, as may be seen from equation (10), formed the magnitude F by simply integrating according to v the expression given for the differential coefficient according to v. In order to see in what way the magnitude $\frac{1}{A}Z$ differs from this, we will modify somewhat the expression for Z given in the preceding section.

According to equation (2),

$$\frac{T}{A} dZ = dL.$$

dL denotes here the internal and external work, taken together, which is performed when the body undergoes an infinitely slight change of condition. We will denote the internal work by dI; and since, when the condition of the body is determined by its temperature T and its volume v, I must be a function of these two quantities, we may write

$$dI=\frac{dI}{dT}dT+\frac{dI}{dv}dv.$$

The external work, assuming it to consist merely in overcoming an external pressure, is represented by pdv. Hence, if we further decompose the differential dZ into its two parts, we may write the above equation thus:—

$$\frac{T}{A}\frac{dZ}{dT}dT+\frac{T}{A}\frac{dZ}{dv}dv=\frac{dI}{dT}dT+\left(\frac{dI}{dv}+p\right)dv,$$

whence we have

$$\left.\begin{aligned}\frac{T}{A}\frac{dZ}{dT}&=\frac{dI}{dT},\\ \frac{T}{A}\frac{dZ}{dv}&=\frac{dI}{dv}+p.\end{aligned}\right\} \quad . \quad . \quad . \quad . \quad . \quad . \quad (13)$$

Differentiating the first of these equations according to v, and the second according to T, we get

$$\frac{T}{A}\frac{d^2Z}{dTdv}=\frac{d^2I}{dTdv},$$

$$\frac{1}{A}\frac{dZ}{dv}+\frac{T}{A}\frac{d^2Z}{dTdv}=\frac{d^2I}{dTdv}=\frac{dp}{dT}.$$

The first of these equations subtracted from the second, gives

$$\frac{1}{A}\frac{dZ}{dv}=\frac{dp}{dT}.$$

The differential coefficient of Z according to v consequently fulfils the condition given in (12); the second of the equations (13) gives at the same time the differential coefficient according to T; and putting these two together, we obtain the complete differential equation

$$\frac{1}{A}dZ=\frac{1}{T}\frac{dI}{dT}dT+\frac{dp}{dT}dv. \quad . \quad . \quad . \quad . \quad (14)$$

To obtain the quantity $\frac{1}{A}Z$, we must integrate this equation. It is easy to see that this integral will in general differ by a function of T from that which would be obtained by integrating only the last term. The two integrals can only be regarded as directly equal if $\frac{dI}{dT}=0$, whence also, in order that the foregoing equation may be integrable, it follows that $\frac{d^2p}{dT^2}=0$, a case which occurs in perfect gases.

I believe that what I can claim as new in my equation (II.) is just this, that the magnitude Z which there occurs has acquired, through my developments, a definite physical meaning, whence it follows that it is fully determined by the arrangement of the constituent molecules of the body existing at any given instant. Thus only does it become possible to deduce from this equation the important conclusion which follows.

[To be continued.]

XXIX. *On the Application of the Theorem of the Equivalence of Transformations to the Internal Work of a mass of Matter.* By Professor R. Clausius.

[Concluded from p. 97.]

§ 7. We will now investigate the manner in which, from equation (II.), it is possible to arrive at the equation (I.) previously given in § 1, which equation must apply, according to the fundamental theorem that I have already enunciated, to every *reversible circular process.*

When successive alterations of condition constitute a *circular process*, the disgregation of the body is the same at the end of the operation as it was at the beginning, and hence the following equation must hold good:—

$$\int dZ=0. \quad . \quad . \quad . \quad . \quad . \quad . \quad . \quad . \quad (15)$$

Equation (II.) is hereby transformed into

$$\int\frac{dQ+dH}{T}=0. \quad . \quad . \quad . \quad . \quad . \quad . \quad (16)$$

In order that this equation may accord with equation (I.), namely,

$$\int\frac{dQ}{T}=0,$$

the following equation must be applicable to every reversible circular process:—

$$\int\frac{dH}{T}=0. \quad . \quad . \quad . \quad . \quad . \quad . \quad . \quad (III.)$$

It is this equation which leads to the consequences referred to in the introduction as at variance with commonly received views. It can, in fact, be proved that, in order that this equation may be true, it is at once *necessary* and *sufficient* to assume the following theorem:—

The quantity of heat actually present in a body depends only on its temperature, and not on the arrangement of its component particles.

It is at once evident that the assumption of this theorem *suf-*

fices for equation (III.); for if H is a function of the temperature only, the differential expression $\frac{dH}{T}$ takes the form $f(T)dT$, in which $f(T)$ is obviously a real function which can have but one meaning, and the integral of this expression must plainly be equal to nothing if the initial and final values of T are the same.

The *necessity* of this theorem may be demonstrated thus.

In order to be able to refer the alterations of condition to alterations of certain magnitudes, we will assume that the manner in which the body changes its condition is not altogether arbitrary, but is subject to such conditions that the condition of the body is determined by its temperature, and by any second magnitude which is independent of the temperature. This second magnitude must plainly be connected with the arrangement of the constituent particles: we may, for example, consider the disgregation of the body as such a magnitude; it may, however, be any other magnitude dependent on the arrangement of the constituent particles. A case which often occurs, and one which has been frequently discussed, is that in which the *volume* of the body is the second magnitude, which can be altered independently of the temperature, and which, together with the temperature, determines the condition of the body. We will take X as a general expression for the second magnitude, so that the two magnitudes T and X together determine the condition of the body.

Since, however, the quantity of heat, H, present in the body is a magnitude which in any case is completely determined by the condition of the body at any instant, it must here, where the condition of the body is determined by the magnitudes T and X, be a function of these two magnitudes. Accordingly, we may write the differential dH in the following form,

$$dH = MdT + NdX, \quad . \quad . \quad . \quad . \quad . \quad . \quad (17)$$

where M and N are functions of T and X, which must satisfy the well-known equation of condition to which the differential coefficients of a function of two independent variables are subject; that is, the equation

$$\frac{dM}{dX} = \frac{dN}{dT}. \quad . \quad . \quad . \quad . \quad . \quad . \quad . \quad (18)$$

Again, if the integral $\int \frac{dH}{T}$ is to become equal to nothing each time that the magnitudes T and X return to the same values as they had at the beginning, $\frac{dH}{T}$ must also be the complete differential of a function of T and X. And since we may write, as a consequence of (17),

$$\frac{dH}{T}=\frac{M}{T}dT+\frac{N}{T}dX, \quad . \quad . \quad . \quad . \quad . \quad . \quad (19)$$

we obtain, for the differential coefficients which here occur, the equation of condition

$$\frac{d}{dX}\left(\frac{M}{T}\right)=\frac{d}{dT}\left(\frac{N}{T}\right), \quad . \quad . \quad . \quad . \quad . \quad . \quad (20)$$

which exactly corresponds to equation (18).

By carrying out the differentiations, this equation becomes

$$\frac{1}{T}\frac{dM}{dX}=\frac{1}{T}\frac{dN}{dT}-\frac{N}{T^2}; \quad . \quad . \quad . \quad . \quad . \quad . \quad (21)$$

and, by applying equation (18) to this, we get

$$N=0. \quad . \quad . \quad . \quad . \quad . \quad . \quad . \quad . \quad (22)$$

According to (17), N is the differential coefficient of H according to X; and if this differential coefficient is to be generally equal to nothing, H itself must be independent of X; and since we may understand by X any magnitude whatever which is independent of T, and together with T determines the condition of the body, it follows that H can only be a function of T.

§ 8. This last conclusion appears, according to commonly received opinions, to be opposed to well-known facts.

I will choose as an illustrative example, in the first place, a case which is very familiar, and in which the discrepancy is particularly great, namely, water in its various states. We may have water in the liquid state, and in the solid state in the form of ice, at the same temperature; and the above theorem asserts that the quantity of heat contained in it is in both cases the same. This appears to be contradicted by experience. The specific heat of ice is only about half as great as that of liquid water, and this appears to furnish grounds for the following conclusion. If at any given temperature a unit of weight of ice and a unit of weight of water in reality contained the same quantity of heat, we must, in order to heat or cool them both, impart to or withdraw from the water more heat than we impart to or withdraw from the ice, so that the equality in the quantity of heat could not be maintained at any other temperatures. A similar difference to that existing between water and ice also exists between water and steam, inasmuch as the specific heat of steam is much smaller than that of water.

To explain this difference, I must recall the fact that only part of the heat which a body takes up when heated goes to increase the quantity of heat actually present in it, the remainder being consumed as work. I believe now that the differences in the specific heat of water in its three states of aggregation is caused by great differences in the proportion which is consumed as work, and that this proportion is considerably greater in the liquid state than in the other two states*. We must, accordingly, here distinguish between the observed specific heat and the true specific heat with which the alteration of temperature dT must be multiplied, in order that we may obtain the corresponding increase of the quantity of heat actually present; and, in accordance with the above theorem, I believe we must admit that the true specific heat of water is the same in all three states of aggregation; and the same considerations as apply to water must naturally also apply in like manner to other substances. In order to determine experimentally the true specific heat of a substance, it must be taken in the form of strongly overheated vapour, in such a state of expansion, in fact, that the vapour may, without sensible error, be regarded as a perfect gas; and its specific heat must then be determined under constant volume.

Rankine is not of my opinion in relation to the specific heat of bodies in different states of aggregation. At page 307 of his 'Manual of the Steam-Engine,' he says, "The real specific heat of each substance is constant at all densities, so long as the substance retains the same condition, solid, liquid, or gaseous; but a change of real specific heat, sometimes considerable, often accompanies the change between any two of these conditions." In the case of water in particular, he says, on the same page, that the true specific heat of liquid water is "sensibly equal" to the apparent specific heat; whereas, according to the view above put forth by myself, it must amount to less than half the apparent specific heat.

If Rankine admits that the true specific heat may be different in different states of aggregation, I do not see what reason there

* I have already enunciated this view in my first memoir on the Mechanical Theory of Heat, having, in fact, inserted the following in a note (Poggendorff's *Annalen*, vol. lxxix. p. 376), which has reference to the diminution of the cohesion of water with increase of temperature:—"Thence it follows, also, that only part of the quantity of heat which water receives from without when heated, is to be regarded as heat in the free state, while the rest is consumed in diminishing the cohesion. In accordance with this view is the circumstance that water has so much higher a specific heat than ice, and probably also than steam." At that time the experiments of Regnault on the specific heat of the gases were not yet published, and we still found in the text-books the number 0·847, obtained by De la Roche and Bérard, for the specific heat of steam. I had, however, already concluded, from the theoretical grounds which are the subject of the present discussion, that this number must be much too high; and it is to this conclusion that the concluding words, "and probably also than steam," refer.

is for supposing it to remain constant within the same state of aggregation. Within one and the same state of aggregation, *e. g.* within the solid state, alterations in the arrangement of the molecules occur, which, though without doubt less considerable, are still essentially of the same kind as the alterations which accompany the passage from one state of aggregation to another; and it therefore seems to me that there is something arbitrary in denying for the smaller alterations what is admitted in respect to the greater. On this point I cannot agree with the way in which the talented English mathematician treats the subject; but, relying simply on the theorem established by myself in relation to the working force of heat, it appears to me that only one of the following cases can be possible. Either this theorem is correct, in which case the true specific heat remains the same, not only for the same state of aggregation, but for the different states of aggregation, or the theorem is not correct, and in this case we have no definite knowledge whatever concerning the true specific heat, and it may equally well be variable within the same state of aggregation as in different states of aggregation.

§ 9. I believe, indeed, that we must extend the application of this theorem, supposing it to be correct, still further, and especially to *chemical combinations and decompositions.*

The separation of chemically combined substances is likewise an increase of the disgregation, and the chemical combination of previously isolated substances is a diminution of their disgregation; and consequently these processes may be brought under considerations of the same class as the formation or precipitation of vapour. That in this case also the effect of heat is to increase the disgregation, results from many well-known phenomena, many compounds being decomposible by heat into their constituents—as, for example, mercuric oxide, and, at a very high temperature, even water. To this it might perhaps be objected that, in other cases, the effect of increased temperature is to favour the union of two substances—that, for instance, hydrogen and oxygen do not combine at low temperatures, but do so easily at higher temperatures. I believe, however, that the heat exerts here only a secondary influence, contributing to bring the atoms into such relative positions that their inherent forces, by virtue of which they strive to unite, are able to come into operation. Heat itself can never, in my opinion, tend to produce combination, but only, and in every case, decomposition.

Another circumstance which renders the consideration of this case more difficult is this, that the conclusions we have been accustomed to draw always imply that the alterations in question can take place in a constant and reversible manner; this, however, is not usually the case under the circumstances which accompany our chemical operations. Nevertheless cases do occur in which this condition is fulfilled, especially in the chemical changes brought about by the action of electric force. The galvanic current affords us a simple means of causing combination or decomposition; and in this case the cell in which the chemical change takes place itself forms a galvanic element, the electromotive force of which either contributes to intensify the current, or has to be overcome by other electromotive force; so that in the one case there is a production, and in the other a consumption of work.

Similarly, I believe that we could in all cases, by producing or expending work, cause the combination or separation of substances at pleasure, provided we possessed the means of acting at will on the individual atoms, and of bringing them into whatever position we pleased. At the same time I am of opinion that heat, leaving out of view its secondary effects, tends in a definite manner, in all cases of chemical change, to render the combination of atoms more difficult, and to facilitate their separation, and that the energy of its action is likewise regulated by the general law above given.

Supposing this to be the case, the theorem which we have deduced from this law must also be applicable here, and a chemical compound must contain exactly the same quantity of heat as its constituents would contain at the same temperature in the uncombined state. Hence it follows that the true specific heat of every compound must admit of being simply calculated from the specific heats of the simple bodies. If we further take into consideration the well-known relation between the specific heats of the simple bodies and their atomic weights (a relation which I believe not only to be nearly, but, in the case of the true specific heats, absolutely exact), it is apparent what enormous simplifications the law which we have established is capable, supposing it to be true, of introducing into the doctrine of heat.

§ 10. After these expository remarks, I can now cite the more extended form of the theorem of the equivalence of transformations.

In § 1 I have mentioned two kinds of transformations: first, the transformation of work into heat, and *vice versâ*; and secondly, the transference of heat between bodies of different temperatures. In addition to these, we will now take, as a third kind of transformation, the alteration of the disgregation of a body, assuming the increase of disgregation as a positive, and the diminution of it as a negative, transformation.

We will now, in the first place, bring the first and last transformation into relation with each other; and here the same circumstances have to be taken into consideration as have already been discussed in § 5. If a body changes its disgregation in a reversible manner, the change is accompanied by a transformation of heat into work, or of work into heat, and we can determine the equivalent values of the two kinds of transformations by comparing together the transformations which take place simultaneously.

Let us first assume that *a constant alteration of arrangement takes place at different temperatures*; the quantity of heat which is thereby converted into work, or is produced from work, is then different in the different cases, and is, in fact, according to the above law, proportional to the absolute temperature. If, now, we regard as equivalent the transformations which correspond to one and the same alteration of arrangement, it results that, for the determination of the equivalent values of these transformations, we must divide the several quantities by the absolute temperatures respectively corresponding to them. The production of the quantity of heat Q from work must, therefore, if it takes place at the temperature T, have the equivalent value

$$\frac{Q}{T} \text{ const.};$$

and if we here take the constant, which can be assumed at will, as equal to unity, we obtain the expression given in § 1.

We will assume, in the second place, that *various alterations of arrangement occur at a constant temperature*, these alterations being accompanied by increase of disgregation; and if we adopt as a principle that increments of disgregation wherein the same quantity of heat is converted into work shall be regarded as equivalent to each other, and that their equivalent value shall be equal, when taken absolutely, to that of the simultaneously occurring transformation from heat into work, but that they shall have the opposite sign, we thus acquire a starting-point for the determination of the equivalent values of alterations of disgregation.

By combining these two rules, we can determine also the equivalent value of an alteration of disgregation occurring at various temperatures, and we thus obtain the expression given in § 5. Let, for instance, dL be an element of the work performed during an alteration of disgregation, in effecting which the quantity of heat AdL is consumed, and let the equivalent value of the alteration of disgregation be denoted by $Z-Z_0$, we then have

$$Z-Z_0=A\int\frac{dL}{T}.$$

Finally, as to the process cited above as the second kind of transformation—namely, the passage of heat between bodies of different temperatures,—in the case of reversible alterations of condition it can be brought about only by heat being converted into work at the one temperature, and work back again into heat at the other; it is therefore already comprised among the transformations of the first kind. And, as I have mentioned in my previous memoir, we may in all cases regard a transformation of the second kind as a combination of two transformations of the first kind.

We will now return to equation (II.), namely,

$$\int \frac{dQ+dH}{T} + \int dZ = 0.$$

dH is here the increment received by the quantity of heat present in the body during an infinitely small change of condition, and dQ is the quantity of heat simultaneously given up to external bodies. The sum $dQ+dH$ is therefore the quantity of heat which, supposing it to be positive, is freshly produced from work, or if it is negative, must be converted into work. Accordingly, the first integral in the above equation is the equivalent value of all the transformations which have occurred of the first kind; the second integral represents the transformations of the third kind; and the sum of all these transformations must be, as is expressed by the equation, equal to nothing.

Hence, in so far as it regards *reversible* alterations of condition, the theorem may be expressed in the following form:—

If the equivalent value $\frac{Q}{T}$ *be assumed for the production of the quantity of heat* Q *from work at the temperature* T, *a magnitude admits of being introduced, as the second transformation corresponding thereto, which has relation to the alterations of the condition of the body, and is completely determined by the initial and final conditions of the body, and which fulfils the condition that in every reversible alteration of condition the algebraic sum of the transformations is equal to nothing.*

§ 11. We must now examine the manner in which the foregoing theorem is modified when we give up the condition that all alterations of condition are to take place reversibly.

From what has been said in § 4 concerning non-reversible alterations of condition, it is easy to perceive that the following must be the general behaviour of all three kinds of transformations. A negative transformation can never occur without a simultaneous positive transformation whose equivalent value is at least as great; on the other hand, positive transformations are not necessarily accompanied by negative transformations of equal value, but may take place in conjunction with smaller negative transformations, or even without any at all.

If heat is to be transformed into work, which is a negative transformation, a positive alteration of disgregation must take place at the same time, which cannot be smaller in amount than that determinate magnitude which we regard as equivalent. In the positive transformation of work into heat, on the other hand, the state of things is different. If the force of heat is overcome by opposed forces, so that a negative change of disgregation is brought about, we know that in this case the overcoming forces may be greater than is required to produce the particular result. The excess of force may then give rise to motions of considerable velocity in the particles of the body under consideration, and these motions may subsequently be changed into the molecular motions which we call heat, so that in the end more work comes to be transformed into heat than corresponds to the negative change of disgregation brought about. In many operations, especially in friction, the transformation of work into heat may take place even quite independently of any simultaneous negative transformation.

The relation in which the third kind of transformation, namely alteration of disgregation, stands to considerations of this nature, is implied in what has been already said. The positive alteration of disgregation may indeed be greater, but cannot be smaller, than the accompanying transformation of heat into work; and the negative alteration of disgregation may be smaller, but cannot be greater, than the transformation of work into heat.

Finally, in so far as regards the second kind of transformation, or the passage of heat between bodies of different temperatures, I have thought myself justified in assuming as a fundamental proposition what, according to all that we know of heat, must be regarded as self-evident, namely, that the passage from a lower to a higher temperature, which counts as a negative transformation, cannot take place of itself—that is, without a simultaneous positive transformation. On the other hand, the passage of heat in the contrary direction, from a higher to a lower temperature, may very well take place without a simultaneous negative transformation.

Taking these circumstances into consideration, we will now return once more to the consideration of the development by means of which we arrived at equation (II.) in § 5. Equation (2), which occurs in the same section, expresses the relation in which an infinitely small alteration of disgregation must stand to the work simultaneously performed by the heat, under the condition that the alteration takes place in a reversible manner. In case this last condition need not be fulfilled, the alteration of disgregation may be greater, provided it is positive, than the value calculated from the work; and if negative, it may be, when taken absolutely, smaller than that value, but in this case also it would algebraically have to be stated as higher. Instead of equation (2), we must therefore write

$$dZ \geq \frac{A\,dL}{T} \quad . \quad . \quad . \quad . \quad . \quad . \quad . \quad . \quad (2a)$$

Applying this to equation (1), we obtain, instead of equation (5),

$$\frac{dQ + dH}{T} + dZ \geq 0. \quad . \quad . \quad . \quad . \quad (5a)$$

The further question now arises, what influence would it have on the formulæ, if a direct passage of heat took place between parts of different temperature within the body in question.

In case the body is not of uniform temperature throughout, the differential expression occurring in equation (5a) must not be referred to the entire body, but only to a portion whose temperature may be considered as the same throughout; so that if the temperature of the body varies continuously, the number of parts must be assumed as infinite. In integrating, the expressions which apply to the separate parts may be united again to a single expression for the whole body, by extending the integral, not only to the alterations of one part, but to the alterations of all the parts. In forming this integral, we must now have regard to the passage of heat taking place between the different parts.

It must here be remarked that dQ is an element of the heat which the body under consideration gives up to, or absorbs from, an external body which serves only as a reservoir of heat, and that this element does not come into question now that we are discussing the passage of heat between the different parts of the body itself. This transfer of heat is mathematically expressed by a decrease in the quantity of heat H in one part, and an equivalent increase in another part; and accordingly we require to direct our attention only to the term $\frac{dH}{T}$ in the differential expression (5a). If we now suppose that the infinitely small quantity of heat dH leaves one part of the body whose temperature is T_1, and passes into another part whose temperature is T_2, there result the two following infinitely small terms,

$$-\frac{dH}{T_1} \text{ and } +\frac{dH}{T_2},$$

which must be contained in the integral; and since T_1 must be greater than T_2, it follows that the positive term must in any case be greater than the negative term, and that consequently the algebraic sum of both is positive. The same thing applies equally to every other element of heat transferred from one part to another; and the alteration which the integral of the whole differential expression occurring in (5*a*) undergoes, on account of this transfer of heat, can therefore only consist in the addition of a positive quantity to the value which would else have been obtained. But since, as results from equation (5*a*), the first value which would be obtained, without taking this direct transfer of heat into consideration, cannot be less than nothing, this can still less be the case when it has been increased by another positive quantity.

We may therefore write as a general expression, including all the circumstances which occur in non-reversible alterations, the following, instead of equation (II.):—

$$\int \frac{dQ + dH}{T} + \int dZ \geqq 0. \quad . \quad . \quad . \quad . \quad (IIa)$$

The theorem which in § 1 was enunciated in reference to circular processes only, and was represented by the expression (I*a*), has thus assumed a more general form, and may be enunciated thus:—

The algebraic sum of all the transformations occurring in any alteration of condition whatever can only be positive, or, as an extreme case, equal to nothing.

In my previous paper I have spoken of two transformations with opposite signs, which neutralize each other in the algebraic sum, as *compensating* transformations. The foregoing theorem may therefore be enunciated still more briefly as follows:—

Uncompensated transformations can only be positive.

§ 12. In conclusion, we will submit the integral

$$\int \frac{dH}{T},$$

which has been frequently used above, to a somewhat closer consideration. We will call this integral, when it is taken from any given initial condition to the condition actually existing, *the equivalent value of the heat in the body* (Körperwärme) *calculated from the given initial condition.* That is, when in any way whatever work is transformed into heat, or heat into work, and the quantity of heat present in the body thereby altered, the increment or decrement of this integral gives the equivalent value of the transformations which have taken place. Further, if transfers of heat take place between parts of different temperature within the body itself, or within a system of bodies, the equivalent value of these transfers of heat is likewise expressed by the increment or decrement of this integral, if it is extended to the whole system of bodies under consideration.

In order to be able actually to perform the integration which has been indicated, we must know the relation between the quantity of heat H and the temperature T. If we call the mass of the body m, and its true specific heat c, we have, for an alteration of temperature throughout amounting to dT, the equation

$$dH = mcdT. \quad \quad (23)$$

According to what has been said above, the true specific heat of a body is independent of the arrangement of its particles; and since a condition is known, namely, that of the perfect gases, for which we must regard it as established, partly by existing experimental data, and partly as the result of theoretical considerations, that the true specific heat is independent of temperature, we may assume the same thing for the other states of aggregation, and may regard the true specific heat as always *constant*. Thence it follows that the amount of heat present in a body is simply proportional to its absolute temperature, inasmuch as we can write

$$H = mcT. \quad \quad (24)$$

The foregoing equation still remains applicable even when the body is not homogeneous, but consists of different substances, all, however, at the temperature T, if for c we substitute the corresponding mean value. On the other hand, if different parts of the body have different temperatures, we must in the first instance apply the equation to the separate parts, and then unite the various equations by summation. If, for the sake of generality, we assume that the temperature varies continuously, so that the body must be conceived as divided into an infinite number of parts, the equation takes the following form:

$$H = \int cTdm. \quad \quad (25)$$

Applying this expression to the integral given above for the transformation-value of the heat in the body, and denoting the initial temperature by T_0, we obtain, for the more simple case in which the temperature is uniform throughout,

$$\int \frac{dH}{T} = mc \int_{T_0}^{T} \frac{dT}{T} = mc \log \frac{T}{T_0}, \quad \quad (26)$$

and, as a general expression embracing all cases,

$$\int \frac{dH}{T} = \int c \log \frac{T}{T_0} \cdot dm. \quad \quad (27)$$

If the disgregation of a body is altered, without heat being supplied to or withdrawn from it, by an external object, the amount of heat contained in the body must be changed in consequence of the production or consumption of heat attendant on the alteration of disgregation, and a rise or fall of temperature must be the result; consequently the question may be raised, How great must the alteration of disgregation be in order to bring about a given change of temperature, it being assumed that all alterations of condition take place reversibly? In this case we must apply equation (II.), putting $dQ=0$, whereby it is transformed into

$$\int \frac{dH}{T} + \int dZ = 0. \quad . \quad . \quad . \quad . \quad . \quad . \quad 28)$$

If we assume, for the sake of simplicity, that the temperature of the entire body varies uniformly, so that T has the same value for all parts, we may apply equation (26) to the determination of the first of the two integrals; and we thus obtain, for the alteration of disgregation sought, the equation

$$Z - Z_0 = mc \log \frac{T_0}{T}. \quad . \quad . \quad . \quad . \quad . \quad (29)$$

If we desired to cool a body down to the absolute zero of temperature, the corresponding alteration of disgregation, as shown by the foregoing formula, in which we should then have $T=0$, would be infinitely great. Herein lies a chief argument for supposing it to be impossible to produce such a degree of cold, by any alteration of the condition of a body, as to arrive at the absolute zero.

8

ON DIFFERENT FORMS OF THE FUNDAMENTAL EQUATIONS OF THE MECHANICAL THEORY OF HEAT AND THEIR CONVENIENCE FOR APPLICATION

Rudolf Clausius

This article was translated expressly for this Benchmark volume by R. B. Lindsay, Brown University, from "Über verschiedene für die Anwendung bequeme Formen der Hauptgleichungen der mechanischen Wärmetheorie," in Abhandlungen über die mechanische Wärmetheorie, *Vol. II, Vieweg, Braunschweig, 1867, pp. 1–44*

Presented to the Züricher naturforschende Gesellschaft on April 24, 1865. Printed in the *Quarterly Journal of the Gesellschaft*, Vol. 10, p. 1. *Poggendorfs Annalen*, Vol. 125, p. 313, 1865. *Journal de Liouville*, 2d ser., Vol. 10, p. 361.[1]

In my previous publications on the mechanical theory of heat I have had as my purpose the provision of a sound basis for the theory by seeking to put the second law in its simplest and most general form and to demonstrate its necessity. It is more difficult to understand than the first law. I have discussed specific applications only insofar as they appeared to be useful explanatory examples or of value in practice.

As the principles of the mechanical theory of heat came to be recognized as correct, there naturally developed in physical and mechanical circles the desire to apply them to different kinds of phenomena: Since the appropriate differential equations must be handled rather differently than the more usual variety of superficially similar form, this has led to difficulty in many calculations, putting obstacles in the way of mathematical development or even leading to errors. Under these circumstances I have thought I might render a service to workers in physics and mechanics by beginning with the equations in their most general form and then transforming them into forms suitable for special cases. They can then be applied immediately, proving more convenient than in the more general form.

1. The whole mechanical theory of heat rests on two fundamental principles, namely, the law of the equivalence of heat and work and the law of the equivalence of transformations.

In order to express the first principle analytically we consider an arbitrary body

[1] This article, which was first published after the appearance of Part I of the author's "Abhandlungensammlung," is most closely related to the articles in that collection, and I am permitting it to form Part II so that it may follow the first part as closely as possible.

that changes its state and consider the quantity of heat that must be communicated to it during this change of state. Let us designate this quantity of heat by the symbol Q, it being understood that a quantity of heat given up by the body is to be counted as a *negative* quantity communicated *to* the body. Then for an infinitely small change of state for which the corresponding quantity of heat communicated to the body is represented by dQ, the following equation holds:

$$dQ = dU + A\,dW. \tag{1}$$

Here U denotes the quantity that I first introduced into the theory of heat in my 1850 article and defined as the sum of the free heat and the heat consumed in internal work.[2] W. Thomson later proposed the name *energy* of the body for the quantity U.[3] I have adopted his terminology as a very useful one. I have indeed introduced the reservation that in cases in which both components of U should be individually specified, we should also use for it the term *heat and work content*, reproducing my original definition in somewhat simpler form. W denotes the external work done by the body during its change of state, and A is the heat equivalent of the unit of work, or in brief, the calorific equivalent of work [The reciprocal of what later came to be known as the mechanical equivalent of heat.—Ed.]. Hence AW is the external work measured in heat units or simply the external work,[4] in accordance with a simplified mode of expression proposed by me.

If for brevity we denote the external work by a single letter, setting

$$AW = w,$$

we can write the former equation in the form

$$dQ = dU + dw. \tag{1a}$$

In order to express the second principle analytically in its simplest form, we shall assume that the changes undergone by the body form a cyclic process in which the body ultimately returns to its original state. Let dQ again denote an element of heat taken up by the body and let T denote the temperature of the body (measured from the absolute zero of temperature) that prevails at the moment when the body takes up this element of heat. In case the body has several different temperatures in its various parts, T will denote the temperature of the part that takes up the heat dQ. If we then divide the element of heat by the absolute temperature corresponding to it and integrate the differential expression thus obtained over the whole cycle, the following relation holds:

$$\int \frac{dQ}{T} \leqslant 0. \tag{2}$$

[2] *Pogg. Ann.* Vol. 79, p. 385. In this collection, Article 1, p. 33. [Paper 5 of this book.—Ed.]
[3] *Phil. Mag.* 4th Series, Vol. 9, p. 523.
[4] See Appendix A to Article VI. [Reprinted as Paper 7 in this book; the appendix, of later origin, is not reproduced here.—Ed.]

Here the equality sign is to be used when all the changes making up the cyclic process are reversible. If the changes are not reversible, the inequality sign prevails.[5]

2. At first we consider more closely the quantities occurring in Eq. (1a) with respect to their behavior for different kinds of changes of the body.

The external work done where a body passes from an initial state to a definite final state does not depend sloely on the initial and final states but depends also on the nature of the transition.

In the first place, we have to consider whether the external forces acting on the body that are either outweighed by the oppositely acting forces of the body itself or conversely outweight the latter (in which cases the external work is positive or negative respectively), or whether the body's own forces at each instant are equal to or different from the impressed forces, in which case differences can exist only if the overcoming force is greater than that overcome. We can indeed say that every time a force is to outweight another it must be greater than the latter. Since, however, the difference between them can be arbitrarily small we can consider the case in which absolute equality exists as the limiting case–which, though it is never attained in reality, is theoretically possible. If force and oppositely directed counterforce are different, the change taking place is irreversible.

In the second place, if it is determined that the change shall take place reversibly, the external work also depends on the intervening states the body passes through in going from its initial to its final state, or to use a more pictorial expression, on the path of the body from initial to final state.

The energy U of the body, whose elementary change appears in Eq. (1a) along with the element of external work, behaves in quite a different fashion. For given initial and final states, the change in the energy is completely determined by them without the need of knowing how the transition from the one state to the other has taken place. Neither the path followed by the transition nor the circumstance whether the transition takes place reversibly or irreversibly has any influence on the change in energy. Accordingly, if the initial state and the energy corresponding to it are given, we can say that the energy is completely determined by the instantaneous state of the body.

With regard, finally, to the heat Q taken up by the body during the change in state, we can say that since it is the sum of the change in energy and the external

[5] In my article "On an Alternative Form of the Second Principle of the Mechanical Theory of Heat" (Article IV of this collection), in which I first gave the general analytical expression of the second principle for cyclic processes, I chose the sign of the differential dQ otherwise than here: The element of heat given up by a body in its changes was considered positive and the heat communicated to it by a reservoir was taken to be negative. With this choice of sign, which is convenient from certain theoretical points of view, Eq. (2) should be written

$$\int \frac{dQ}{T} \geqslant 0$$

In the present article, however, we shall use the sign terminology in accordance with which Eq. (2) holds.

work that is performed, it must depend on the way in which the transition from the one state to the other takes place in the same fashion as the external work.

In order to delimit the domain of discussion to be handled at first, in what follows it will always be assumed that we are dealing throughout with reversible processes, unless it is specifically stated that irreversible processes are included in the investigations.

Equation (1a), which expresses the content of the first law, holds for both irreversible and reversible changes, and hence to apply it to reversible processes it is not necessary to modify it in any way except by making sure that the external work and quantity of heat denoted by w and Q respectively are understood to refer to reversible processes.

If the relation (2), which is the expression of the second principle, is to be applied to reversible processes, we must in the first place understand by Q the quantity of heat relating to reversible processes and secondly, we must use, in place of the double sign $\leqslant$, the equality sign only. Hence for all reversible cyclic processes we have

$$\int \frac{dQ}{T} = 0. \tag{2a}$$

3. In order to make calculations with Eqs. (1a) and (2a) we shall assume that the state of the body in question is determined by arbitrary quantities. Cases commonly met with are those in which the state of the body is determined by its temperature and volume, or by temperature and pressure, or finally by volume and pressure. However, we shall not restrict ourselves here to particular quantities, but we shall at first assume that the state of the body is fixed by two arbitrary quantities, which we shall call x and y. In our analysis we shall consider these as our independent variables. In specific applications we are indeed always free to think of these as any two of the quantities temperature, volume, and pressure.

If the quantities x and y determine the state of the body, the quantity U—the energy—which depends on the instantaneous state, must be represented as a function of these two variables.

The situation is otherwise with the quantities w and Q. The differential coefficients [*Editor's Note*: For simplicity we shall in future refer to these as derivatives] of these quantities, which we designate as follows:

$$\frac{dw}{dx} = m, \quad \frac{dw}{dy} = n, \tag{1}$$

$$\frac{dQ}{dx} = M, \quad \frac{dQ}{dy} = N, \tag{2}$$

are functions of x and y.

If indeed it happens that the variable x is to become $x + dx$, whereas y remains unchanged, and that the corresponding change of state is to take place reversibly, we are then dealing with a completely determined process and the external work must then also be completely determined. It follows further that the derivative

dw/dx must also have a definite value. A similar situation exists if y goes into $y + dy$, whereas x remains constant. If accordingly the derivatives of the external work w are functions of x and y, in consequence of Eq. (1a) the same will be true of the derivatives of the heat Q taken up by the body, which likewise will be functions of x and y.

Let us now form the expressions for dm and dQ in terms of dx and dy, neglecting quantities of higher order in dx and dy. We obtain

$$dw = M\,dx + n\,dy, \tag{3}$$

$$dQ = M\,dx + N\,dy. \tag{4}$$

We thereby obtain two total differential equations that cannot be integrated as long as the variables x and y are independent of each other. This means that the quantities m, n, M, and N do not satisfy the conditions of integrability, namely,

$$\frac{dw}{dy} = \frac{dn}{dx} \quad \text{and} \quad \frac{dM}{dy} = \frac{dN}{dx}.$$

The quantities w and Q thus belong to the class mentioned in the mathematical introduction to the first part of my collection of memoirs. They have the distinctive property that although their derivatives are definite functions of the two independent variables, they themselves cannot be represented by such functions but can be determined only if another condition connecting the variables is given, thereby prescribing the path pursued by the body in its changes.

4. Let us revert to Eq. (1a) and insert there the expressions (3) and (4) and also decompose dU into its two parts relating to dx and dy respectively. In this way we obtain the equation

$$M\,dx = N\,dy = \left(\frac{dU}{dx} + m\right) dx + \left(\frac{dU}{dy} + n\right) dy.$$

Since this equation must remain valid for all arbitrary values of dx and dy, it yields the following equations

$$M = \frac{dU}{dx} + m,$$

$$N = \frac{dU}{dy} + n.$$

If we differentiate the first of these equations by y and the second by x, we get

$$\frac{dM}{dy} = \frac{d^2U}{dx\,dy} + \frac{dm}{dy},$$

$$\frac{dN}{dx} = \frac{d^2U}{dy\,dx} + \frac{dn}{dx}.$$

We now apply to U the theorem that if we have a function of two independent vaiables the order of differentiation is inmaterial, that is, we can write

$$\frac{d^2U}{dx\,dy} = \frac{d^2U}{dy\,dx}.$$

Taking this result into consideration and subtracting the second of the two foregoing equations from the first, we obtain

$$\frac{dM}{dy} - \frac{dN}{dx} = \frac{dm}{dy} - \frac{dn}{dx}. \tag{5}$$

We handle Eq. (2a) in similar fashion. If we replace dQ there by its value from (4) we get

$$\int \left(\frac{M}{T}\,dx + \frac{N}{T}\,dy\right) = 0.$$

If the integral on the left-hand side is to become zero whenever x and y return to their original values, the expression under the integral sign must be the perfect differential of a function of x and y. Hence the above-mentioned condition of integrability must be fulfilled. In this case it has the form

$$\frac{d}{dy}\left(\frac{M}{T}\right) = \frac{d}{dx}\left(\frac{N}{T}\right).$$

If we carry out the differentiations, remembering that the temperature T of the body is also to be considered a function of x and y, we obtain

$$\frac{1}{T}\cdot\frac{dM}{dy} - \frac{M}{T^2}\cdot\frac{dT}{dy} = \frac{1}{T}\cdot\frac{dN}{dx} - \frac{N}{T^2}\cdot\frac{dT}{dx}.$$

We may write this in the alternative form

$$\frac{dM}{dy} - \frac{dN}{dx} = \frac{1}{T}\left(M\frac{dT}{dy} - N\frac{dT}{dx}\right). \tag{6}$$

We shall give a somewhat different form to Eqs. (5) and (6). In order not to have too many symbols in the formulas, we shall replace M and N (which are the abbreviated symbols for the derivatives dQ/dx and dQ/dy respectively) by the derivatives themselves once more. Let us consider further the difference on the right-hand side of (5). If we again write for the quantities m and n the derivatives dw/dx and dw/dy respectively, the difference in question becomes

$$\frac{d}{dy}\left(\frac{dw}{dx}\right) - \frac{d}{dx}\left(\frac{dw}{dy}\right).$$

The quantity represented by this difference is a function of x and y, which it is customary to assume as known, since the external forces acting on the body are observable and from there the external work can be determined. We shall call this difference, which will appear frequently in what follows, the work-difference with respect to xy and introduce for it a special symbol, by writing

$$E_{xy} = \frac{d}{dy}\left(\frac{dw}{dx}\right) - \frac{d}{dx}\left(\frac{dw}{dy}\right). \tag{7}$$

From this way of putting things Eqs. (5) and (6) become

$$\frac{d}{dy}\left(\frac{dQ}{dx}\right) - \frac{d}{dx}\left(\frac{dQ}{dy}\right) = E_{xy}, \tag{8}$$

$$\frac{d}{dy}\left(\frac{dQ}{dx}\right) - \frac{d}{dx}\left(\frac{dQ}{dy}\right) = \frac{1}{T}\left(\frac{dT}{dy}\cdot\frac{dQ}{dx} - \frac{dT}{dx}\cdot\frac{dQ}{dy}\right). \tag{9}$$

These two equations form the analytic expression of the two principles relating to reversible processes for the case in which the state of the body is determined by two arbitrary variables. From these equations we deduce at once a third, which is simpler insofar as it contains only the differential coefficients of the first order of Q; namely,

$$\frac{dT}{dy}\cdot\frac{dQ}{dx} - \frac{dT}{dx}\cdot\frac{dQ}{dy} = TE_{xy} \tag{10}$$

5. The three preceding equations become particularly simple if we choose the temperature of the body as one of the independent variables. If $y = T$ it follows that

$$\frac{dT}{dy} = 1.$$

With respect to dT/dx, in the formula of this it is assumed that where x goes into $x + dx$, the other variable, previously designated by y, remains constant. Since now T itself is this other variable, which is assumed to be constant in the differential coefficient, it follows that the curve to set

$$\frac{dT}{dx} = 0.$$

If we now form the work difference with respect to x, T, this becomes (11)

$$E_{xT} = \frac{d}{dT}\left(\frac{dw}{dx}\right) - \frac{d}{dx}\left(\frac{dw}{dT}\right),$$

and by application of this value Eqs. (8), (9), and (10) become

$$\frac{d}{dT}\left(\frac{dQ}{dx}\right) - \frac{d}{dx}\left(\frac{dQ}{dT}\right) = E_{XT}, \tag{12}$$

$$\frac{d}{dT}\left(\frac{dQ}{dx}\right) - \frac{d}{dx}\left(\frac{dQ}{dT}\right) = \frac{1}{T}\frac{dQ}{dx}, \tag{13}$$

$$\frac{dQ}{dT} = TE_{XT}. \tag{14}$$

If we place the product TE_{XT} in (14) in place of dQ/dx in Eq. (12) and differentiate it by T we get the following simple equation:

$$\frac{d}{dx}\left(\frac{dQ}{dT}\right) = T\frac{dE_{XT}}{dT}. \tag{15}$$

6. Hitherto we have made no particular assumptions about the external forces to which the body is exposed and on which the external work done by the body during changes of the state depend. We shall now examine more closely a case that often comes up, namely, that in which the only external force present or at least the only one that is sufficiently important to deserve attention in the calculation is a pressure acting on the surface of the body, which is everywhere equally strong and directed everywhere normal to the surface.

In this case external work will be done only if the volume of the body changes. If we denote the pressure per unit area of the surface by p, the external work that is done if the volume v increases by dv is (in mechanical units)

$$dW = p\,dv,$$

and the external work (in heat units) is

$$dw = Ap\,dv. \tag{16}$$

If we now agree that the state of the body is determined by two arbitrary variables x and y, the pressure p and the volume v are functions of x and y. Accordingly we can enter the preceding equation in the form

$$dw = Ap\left(\frac{dv}{dx}\,dx + \frac{dv}{dy}\,dy\right),$$

from which follows

$$\left.\begin{aligned} \frac{dw}{dx} &= Ap\frac{dv}{dx}, \\ \frac{dw}{dy} &= Ap\frac{dv}{dy}. \end{aligned}\right\} \tag{17}$$

If we insert these values of dw/dx and dw/dy in the expression given in (7) for E_{xy}, and carry out the second differentiations indicated there, and note that $d^2v/dx\,dy = d^2v/dy\,dx$, we have

$$E_{xy} = A\left(\frac{dp}{dy}\cdot\frac{dy}{dx} - \frac{dp}{dx}\cdot\frac{dv}{dy}\right). \tag{18}$$

We must now apply this value of E_{xy} to Eqs. (8) and (10).

If x and T are the two independent variables, we get, corresponding to the previous equation,

$$D_{xT} = A\left(\frac{dp}{dT}\cdot\frac{dv}{dx} - \frac{dp}{dx}\cdot\frac{dv}{dT}\right), \tag{19}$$

which value we have to use in Eqs. (12), (14), and (15).

The expression given in (18) assumes its simplest form if we choose either the volume or the pressure as one of the independent variables, or if we choose volume and pressure as the two independent variables. As is easily seen for these cases, Eq. (18) takes the form

$$E_{vy} = A\,\frac{dp}{dv}, \tag{20}$$

$$E_{py} = -A\,\frac{dv}{dy}, \tag{21}$$

$$E_{vp} = A. \tag{22}$$

7. Under the circumstances mentioned in the preceding paragraph, in which the external force is a uniform normal surface pressure, it is customary to choose as the independent variables for the determination of the state of the body those mentioned at the end of the paragraph—namely, volume and temperature, pressure and temperature, or volume and pressure. Although the systems of differential equations applying to these three cases are readily derivable from the above more general formulation, I am setting them down here in precise form because of their frequent application. The first case is the one that I have always used in my articles when special cases have been considered.

If v and T are chosen as the independent variables, we have

$$\begin{gathered}\frac{d}{dT}\left(\frac{dQ}{dv}\right) - \frac{d}{dv}\left(\frac{dQ}{dT}\right) = A\,\frac{dp}{dT},\\ \frac{d}{dT}\left(\frac{dQ}{dv}\right) - \frac{d}{dv}\left(\frac{dQ}{dT}\right) = \frac{1}{T}\,\frac{dQ}{dv},\\ \frac{dQ}{dv} = AT\,\frac{dp}{dT}, \quad \frac{d}{dv}\left(\frac{dQ}{dT}\right) = AT\,\frac{d^2p}{dT^2}.\end{gathered} \tag{23}$$

If p and T are chosen as independent variables, we have

$$\frac{d}{dT}\left(\frac{dQ}{dp}\right) - \frac{d}{dp}\left(\frac{dQ}{dT}\right) = -A\,\frac{dv}{dT},$$

$$\frac{d}{dT}\left(\frac{dQ}{dp}\right) - \frac{d}{dp}\left(\frac{dQ}{dT}\right) = \frac{1}{T}\,\frac{dQ}{dp}, \tag{24}$$

$$\frac{dQ}{dp} = -AT\,\frac{dv}{dT}, \qquad \frac{d}{dp}\left(\frac{dQ}{dT}\right) = -AT\,\frac{d^2v}{dT^2}.$$

If v and p are chosen as the independent variables, we have

$$\frac{d}{dp}\left(\frac{dQ}{dv}\right) - \frac{d}{dv}\left(\frac{dQ}{dp}\right) = A,$$

$$\frac{d}{dp}\left(\frac{dQ}{dv}\right) - \frac{d}{dv}\left(\frac{dQ}{dp}\right) = \frac{1}{T}\left(\frac{dT}{dp}\cdot\frac{dQ}{dv} - \frac{dT}{dv}\,\frac{dQ}{dp}\right), \tag{25}$$

$$\frac{dT}{dp}\cdot\frac{dQ}{dv} - \frac{dT}{dv}\cdot\frac{dQ}{dp} = AT.$$

8. Among the cases to which the equations of the preceding paragraph find application, the simplest is that in which a homogeneous body has the same temperature throughout, is subjected to a uniform, normal surface pressure, and can change its volume with change of temperature and pressure without at the same time changing its state of aggregation.

In this case the derivative dQ/dT has a simple physical meaning. If we assume that the weight of the body [*Editor's Note*: The author evidently means the mass here] is unity, the derivative is the specific heat at constant volume or constant pressure depending on which quantity is kept constant during the differentiation.

When in the nature of the case it is necessary to change the independent variables and hence have derivatives that differ from each other only in the fact that the quantities that remain constant in the differentiation are different, it is convenient to indicate this difference by some external symbol in order not to have to specify it continually in words. I shall do this by placing the derivatives in parentheses and treating the quantities to be kept constant as subscripts with a line over them [*Editor's Note*: We here use the subscript without the line over it]. In accordance with this plan the two derivatives that denote the specific heats at constant volume and constant pressure respectively are represented as follows:

$$\left(\frac{dQ}{dT}\right)_v \quad \text{and} \quad \left(\frac{dQ}{dT}\right)_p.$$

We further note that of the three quantities that come into consideration in the determination of the state of the body—namely, temperature, volume and pressure—each will be a function of the two others, and we can therefore form the following six derivatives:

$$\left(\frac{dp}{dT}\right)_v, \quad \left(\frac{dp}{dv}\right)_T, \quad \left(\frac{dv}{dT}\right)_p, \quad \left(\frac{dv}{dp}\right)_T, \quad \left(\frac{dT}{dv}\right)_p, \quad \left(\frac{dT}{dp}\right)_v.$$

In the preceding derivatives we can omit the specific indices if we decide once and for all that of the three quantities T, v, and p, the one that does not appear in the derivative is to be considered constant. However, for the sake of clarity in the following we shall use the subscript indices throughout.

The calculations to be carried out with the foregoing six derivatives are simplified if we establish in advance the relations connecting them.

It is clear, to begin with, that among these derivatives there are always pairs that are reciprocal to each other. If, for example, we take the quantity v as constant, the other two quantities T and p are so related to each other that each of them can be considered a function of the other. Similarly if p is taken to be constant, T and v stand in this functional relation to each other, and finally if T is constant, the same is true of v and p. We can therefore set

$$\frac{1}{(dT/dp)_v} = \left(\frac{dp}{dT}\right)_v,$$

$$\frac{1}{(dT/dv)_p} = \left(\frac{dv}{dT}\right)_p, \tag{26}$$

$$\frac{1}{(dp/dv)_T} = \left(\frac{dv}{dp}\right)_v.$$

In order to obtain the relation among the three pairs of derivatives we shall, as illustrated, consider p as a function of T and v. We then obtain the total differential equation

$$dp = \left(\frac{dp}{dT}\right)_v dT + \left(\frac{dp}{dv}\right)_T dv.$$

If we now wish to apply this equation to the case in which p is constant, we must place

$$dp = 0$$

and

$$dv = \left(\frac{dv}{dT}\right)_p dT$$

whence we obtain

$$0 = \left(\frac{dp}{dT}\right)_v dT + \left(\frac{dp}{dv}\right)_T \cdot \left(\frac{dv}{dT}\right)_p dT.$$

Taking out the common factor dT and dividing by $(dp/dT)_v$ yields

$$\left(\frac{dp}{dv}\right)_T \cdot \left(\frac{dv}{dT}\right)_p \cdot \left(\frac{dT}{dp}\right)_v = -1. \tag{27}$$

With the help of this equation in combination with Eqs. (26), we can represent each of the six derivatives as a product or ratio of two other derivatives.

9. We now revert to a consideration of the heat taken up and given out by the body and denote the specific heat at constant volume by c and the specific heat at constant pressure by C. Treating the mass of the body as unity, we have

$$\left(\frac{dQ}{dT}\right)_v = c, \quad \left(\frac{dQ}{dT}\right)_p = C.$$

From Eqs. (23) and (24), we then have

$$\left(\frac{dQ}{dv}\right)_T = AT\left(\frac{dp}{dT}\right)_v; \quad \left(\frac{dQ}{dp}\right)_T = -AT\left(\frac{dv}{dT}\right)_p.$$

We can then form the following total differential equations:

$$dQ = c\,dT + AT\left(\frac{dp}{dT}\right)_v dv, \tag{28}$$

$$dQ = C\,dT - AT\left(\frac{dv}{dT}\right)_p dp. \tag{29}$$

From the comparison of these two expressions for dQ we can obtain at once the relation between the two specific heats c and C. Starting with the second equation that treats T and p as independent variables we can deduce an equation in which T and v are the independent variables. All we need to do is to consider p a function of T and v and hence write

$$dp = \left(\frac{dp}{dT}\right)_v dt + \left(\frac{dp}{dv}\right)_T dv.$$

If we insert this expression for dp in Eq. (29) the latter becomes

$$dQ = \left[C - AT\left(\frac{dv}{dT}\right)_p \cdot \left(\frac{dp}{dT}\right)_v\right] dT - AT\left(\frac{dv}{dT}\right)_p \cdot \left(\frac{dp}{dv}\right)_T dv.$$

With the help of Eq. (27) we replace in the foregoing the product in the last term on the right with a single derivative and get

$$dQ = \left[C - AT\left(\frac{dv}{dT}\right)_p \cdot \left(\frac{dp}{dT}\right)_v\right] dT + AT\left(\frac{dp}{dT}\right)_v dv.$$

If we compare this expression for dQ with that in Eq. (28) and realize that the factor of dT in both expressions must be the same, we obtain the following relation between the two specific heats:

$$c = C - AT\left(\frac{dv}{dT}\right)_p \cdot \left(\frac{dp}{dT}\right)_v. \tag{30}$$

The derivative $(dv/dT)_p$ represents the thermal expansion of the body and as a rule is to be taken as known. The other derivative $(dp/dT)_v$ is not generally known from direct observation for solids and liquids. However, from (27) we can set

$$\left(\frac{dp}{dT}\right)_v = -\frac{(dv/dT)_p}{(dv/dp)_T}.$$

In the ratio on the right-hand side of the preceding equation the numerator is the coefficient of thermal expansion already mentioned, and the derivative in the denominator, when taken with the negative sign, represents the compressibility, which has been directly measured for a number of liquids and for solids, and can be calculated approximately from the coefficient of elasticity. If we introduce this ratio, Eq. (30) becomes

$$c = C + AT\frac{[(dv/dT)_p]^2}{(dv/dp)_T}. \tag{31}$$

In the application of this equation to numerical calculations we must observe that the unit volume in the derivatives is the cube of the unit of length used in the denominator of the quantity A. For the unit of pressure we must use the pressure (force) that unit weight exerts on unit area. We must express the coefficient of thermal expansion and compressibility in terms of these units.

Since the derivative $(dv/dp)_T$ is always negative, it follows that the specific heat at constant volume is always smaller than the specific heat at constant pressure. The other derivative $(dv/dT)_p$ is, in general, positive. For water at the temperature of maximum density it is zero, and hence for this temperature the two specific heats of water are equal. For all other temperatures, either above or below that for maximum density, $c < C$. Even if at some temperatures $(dv/dT)_p$ should be negative, it occurs as a square in the expression[6] for $c - C$.

[6] As an example of the application of Eq. (31) we shall take water at a definite temperature and calculate the difference between the two specific heats.

According to the observations of Kopp, whose results are set forth in his

From Eqs. (28) and (29) we can readily derive a total differential equation for Q, which is related to p and v as independent variables. We then need only to consider T as a function of p and v and hence to set

$$dT = \left(\frac{dT}{dp}\right)_v dp + \left(\frac{dT}{dv}\right)_p dv.$$

If we substitute this expression for dT in Eq. (29) we get

$$dQ = \left[C\left(\frac{dT}{dp}\right)_v - AT\left(\frac{dv}{dT}\right)_p\right] dp + C\left(\frac{dT}{dv}\right)_p dv$$

$$= \left(\frac{dT}{dp}\right)_v \left[C - AT\left(\frac{dv}{dT}\right)_p \cdot \left(\frac{dp}{dT}\right)_v\right] dp + C\left(\frac{dT}{dv}\right)_p dv.$$

In the expression on the right the difference in the square bracket is equal to c, according to Eq. (30). Hence we can write the equation in the form

---

Lehrbuch der Physikalischen und Theoretischen Chemie (p. 204), if we take unit volume of water at 0°C, the thermal expansion coefficients are as follows:

at 0°C	-0.000061
at 25°C	+0.000025
at 50°C	+0.00045 .

From the observations of Grassi (*Ann. de chimie et de physique*, 3d series, Vol. 31, p. 431) and Kronig (*Journal für Physik des Auslandes*, Vol. 11, p. 129) the compressibility of water has the following values, which represent the volume decrease due to a pressure increase of one atmosphere as a fraction of the volume at the initial pressure.

at 0°C	0.000050
at 25°C	0.000046
at 50°C	0.000044.

We carry out the relevant calculations for a temperature of 25°C. We choose the meter as the unit of length and the kilogram as the unit of mass. The unit of volume is thus the cubic meter, and since a kilogram of water at 4°C has a volume of 0.001 cubic meter, in order to get $(dv/dT)_p$ we must multiply the above expansion coefficients by 0.001. Thus

$$\left(\frac{dv}{dT}\right)_p = 25 \times 10^{-8}.$$

From the preceding considerations the volume assumed by the water at the temperature in question and at the initial pressure (taken as ordinary normal atmospheric pressure) is taken as unity. At 25°C this value is equal to 0.001001 cubic meter. The atmospheric pressure is taken as unit pressure, whereas we must

$$dQ = c\left(\frac{dT}{dp}\right)_v dp + C\left(\frac{dT}{dp}\right)_p dv. \tag{32}$$

10. The three total differential equations (28), (29), and (32) do not satisfy the condition of immediate integrability, which, with respect to the first, follows from the equations established earlier. If we introduce the symbols c and C in the equations corresponding to Eqs. (23) and (24), we have

$$\begin{aligned}\left(\frac{dc}{dv}\right)_T &= AT\left(\frac{d^2p}{dT^2}\right)_v, \\ \left(\frac{dC}{dp}\right)_T &= -AT\left(\frac{d^2v}{dT^2}\right)_p,\end{aligned} \tag{33}$$

whereas the equations that must be satisfied if (28) and (29) are to be integrable are

---

take the pressure of a kilogram on a square meter as the unit of pressure. Hence in these units atmospheric pressure is represented by 10333. Hence for the negative of the compressibility we have

$$\left(\frac{dv}{dp}\right)_T = -\frac{(0.000046)\,(0.001003)}{(10333)}$$

$$= -45 \times 10^{-13}$$

At 25°C we have to set $T = 273 + 25 = 298$ and for A (after Joule) we take 1/424. If we set these values into Eq. (31) we get

$$C - c = \frac{(298)}{424} \cdot \frac{(25)^2 \times 10^{-16}}{45 \times 10^{-13}} = 0.0098.$$

The corresponding values for 0°C and 50°C came out to be

at 0°C $\quad C - c = 0.0005$
at 50°C $\quad C - c = 0.0358.$

If we take for C, the specific heat at constant pressure, the values obtained experimentally by Regnault, we get for the two specific heats the following pairs of values:

at 0°C	$C = 1$
	$C = 0.9995$
at 25°C	$C = 1.0016$
	$C = 0.9918$
at 50°C	$C = 1.0042$
	$C = 0.9684.$

$$\left(\frac{dc}{dv}\right)_T = A\left[T\left(\frac{d^2p}{dT^2}\right)_v + \left(\frac{dp}{dT}\right)_v\right],$$

$$\left(\frac{dC}{dp}\right)_T = -A\left[T\left(\frac{d^2v}{dT^2}\right)_p + \left(\frac{dv}{dT}\right)_p\right].$$

Similarly we draw the conclusion that Eq. (32) is not integrable, which would indeed follow directly from the fact that it is derived from Eqs. (28) and (29).

The three equations ((28), (29), and (32)) belong accordingly to those total differential equations that were mentioned in the introduction to the first article of my collection and that can be integrated only if some other relation exists among the variables, thereby prescribing the path taken in the change of state.

Among the many applications that can be made of Eqs. (28), (29), and (32), I shall introduce only one as an example. We shall assume that the body changes its volume reversibly by change of pressure without having any heat communicated to it or taken from it. We are to determine what volume change will take place under those circumstances through a given change in pressure and how the temperature is altered. More generally we seek to find the equations connecting temperature, volume, and pressure that exist under these circumstances.

These equations are obtained at once if we set $dQ = 0$ in the three equations just mentioned. Eq. (28) then yields

$$c\,dT + AT\left(\frac{dp}{dT}\right)_v = 0.$$

If we divide this equation by dv, the ratio dT/dv that results is the derivative of T with respect to v for this particular case. We shall distinguish this from other derivatives of T with respect to v by putting the symbol Q as a subscript on the derivative. We thus have

$$\left(\frac{dT}{dv}\right)_Q = -\frac{AT}{c}\left(\frac{dp}{dT}\right)_v. \tag{34}$$

From Eq. (29) we get, analogously,

$$\left(\frac{dT}{dp}\right)_Q = \frac{AT}{c}\left(\frac{dv}{dT}\right)_p, \tag{35}$$

and from Eq. (32) we have

$$\left(\frac{dv}{dp}\right)_Q = -\frac{c}{C}\,\frac{(dT/dp)_v}{(dT/dv)_p},$$

which because of (27), can be written

$$\left(\frac{dv}{cp}\right)_Q = \frac{c}{C}\left(\frac{dv}{dp}\right)_T . \tag{36}$$

If we introduce into this equation the expression for c in (31) we obtain

$$\left(\frac{dv}{dp}\right)_Q = \left(\frac{dv}{dT}\right)_p + \frac{AT}{C}\left[\left(\frac{dv}{dT}\right)_p\right]^2 . \tag{37}$$

11. If we apply the equations of the two preceding paragraphs to an ideal gas, they take on simpler and more definite forms.

For this case we must use the law of Mariotte and Gay-Lussac for the relation connecting T, v, and p, that is,

$$pv = RT, \tag{38}$$

in which R is a constant. From this there follows:

$$\begin{aligned} &\left(\frac{dp}{dT}\right)_v = \frac{R}{v}, \quad \left(\frac{dv}{dT}\right)_p = \frac{R}{p}, \\ &\left(\frac{d^2p}{dT^2}\right)_v = 0 , \quad \left(\frac{d^2v}{dT^2}\right)_p = 0 . \end{aligned} \tag{39}$$

Combination of the last two equations with (33) yields

$$\left(\frac{dc}{dv}\right)_T = 0, \quad \left(\frac{dC}{dp}\right)_T = 0. \tag{40}$$

From this it follows that the two specific heats c and C for an ideal gas can be functions of the temperature only. On other grounds, resting on special considerations into which I shall not go here, it may be concluded that the two specific heats are also independent of the temperature and hence are constant. So far as the specific heat at constant pressure is concerned, this result has been confirmed by the experimental investigations of Regnault on the permanent gases.

Applying the first two equations in (39) to Eq. (30), we get the following for the relation between the two specific heats:

$$c = C - AT \cdot \frac{R}{p} \cdot \frac{R}{v},$$

which with the use of (38), becomes

$$c = C - AR. \tag{41}$$

By applying the first two equations in (39) the Eqs. (28), (29), and (32) take the following form:

$$dQ = c\,dT + AR \cdot \frac{T}{v}\,dv,$$

$$dQ = C\,dT - AR \cdot \frac{T}{p}\,dp, \tag{42}$$

$$dQ = \frac{c}{R} v\,dp + \frac{C}{R} p\,dv,$$

in which we can replace the product AR by the difference $C - c$. In my article "On the Moving Force of Heat, etc" and in an appendix in my collection of articles to the article: "On the Application of the Theorem of the Equivalence of Transformations to Internal Work of a Mass of Matter" I have already given many examples of the applications of Eqs. (42), and shall not go further into this matter here.

12. Another case, which is of particular interest because of its numerous applications, is that in which a partial change in the *state of aggregation* is joined with the change of state of the body in question.

We shall assume that one part of the body in question is in one state of aggregation and the rest in another such state. For example, we might imagine that one part of the body is in the liquid state and the other is in the vapor state with density that it assumes when in contact with the liquid. The equations to be set up are to hold if a part of the body is solid and the rest is liquid or if a part is solid and the rest is in the vapor state. For the sake of generality we shall not define the states of aggregation more specifically, but merely refer to them as first and second states of aggregation.

We then imagine that a certain quantity of material is enclosed in a vessel of given volume and that a part is in the first state of aggregation and the rest is in the second state. If the specific volumes are different for the material in the two states of aggregation at the given temperature, the two parts in the different states are not of arbitrary magnitude but are precisely determined. Thus if the part that is in the state of aggregation of large specific volume increases in magnitude, there is then an increase in pressure that the enclosed material exerts on the surrounding walls and that it therefore conversely experiences from these walls. A point is finally reached in which the pressure is so large that it hinders further transfer to this state of aggregation. Where this point has been reached, if the temperature and spatial extent of the enclosing vessel remain constant, the magnitudes of the portions of the two states of aggregation cannot change further. However, if while the temperature remains constant, the spatial content of the vessel increases, the portion of the material that is the state of aggregation with larger specific volume will grow still further at the expense of the other until the previous pressure is attained, and thereby further change is inhibited.

It is in this that consists the specific character that distinguishes this case from others. If indeed we choose the temperature and volume of the material as the two independent variables, the pressure is not a function of these two variables, but a function of temperature alone. We encounter a similar situation if instead of the volume we choose another quantity that likewise can change independently of the temperature and along with the temperature determines the complete state of the body. We take this quantity as the second independent variable. The pressure cannot depend on this quantity either. The two quantities temperature and pressure together cannot in this case be chosen as the two variables serving to determine the state of the body.

In addition to the temperature we shall use an arbitrary quantity x (still left unspecified) as the second independent variable. We then consider the expression given in (19) for the work difference referred to xT, namely,

$$E_{xT} = A\left(\frac{dp}{dT}\cdot\frac{dv}{dx} - \frac{dp}{dx}\cdot\frac{dv}{dT}\right).$$

According to what has gone before, here we have to set $dp/dx = 0$, and hence we have

$$E_{xT} = A\,\frac{dp}{dT}\cdot\frac{dv}{dx}. \tag{43}$$

If we use this result the three equations (12), (13), and (14) become

$$\frac{d}{dT}\left(\frac{dQ}{dx}\right) - \frac{d}{dx}\left(\frac{dQ}{dT}\right) = A\,\frac{dp}{dT}\cdot\frac{dv}{dx}. \tag{44}$$

$$\frac{d}{dT}\left(\frac{dQ}{dx}\right) - \frac{d}{dx}\left(\frac{dQ}{dT}\right) = \frac{1}{T}\,\frac{dQ}{dx}, \tag{45}$$

$$\frac{dQ}{dx} = AT\,\frac{dp}{dT}\cdot\frac{dv}{dx}. \tag{46}$$

13. In order to give the foregoing equations definite form, we shall call the total mass of the material in question M and the part of this that has gone over to the second state of aggregation m, so that $M - m$ is the mass in the first state of aggregation. We shall treat the quantity m as the independent variable that, together with T, fixes the state of the body.

The specific volume of the material in the first state of aggregation (measured with respect to unit mass) will be denoted by σ, and the specific volume of the material in the second state of aggregation will be denoted by s. Both quantities are related to the temperature T and to the pressure corresponding to this temperature and thus, like pressure, are to be considered as functions of the temperature alone. If we further denote the volume of the body as a whole by v, we have

$$\begin{aligned} v &= (M - m)\sigma + ms \\ &= m\,(s - \sigma) + M\sigma. \end{aligned}$$

For the difference $s - \sigma$ we shall use the symbol u, and hence write

$$v = mu + M\sigma, \tag{47}$$

from which follows

$$\frac{dv}{dm} = u. \tag{48}$$

Let us denote by r the quantity of heat that must be given to the body if a unit mass of this is to pass of temperature T and corresponding pressure from the first state of aggregation to the second. Then

$$\frac{dQ}{dm} = r. \tag{49}$$

We now wish to introduce into the equations the specific heat of the material in the two states of aggregation. The specific heat in question here, however, is neither the specific heat at constant volume nor that at constant pressure. It is given by the quantity of heat needed to heat the body if, simultaneously with the temperature, the pressure changes in the way brought about by the circumstances prevailing for the case in hand. This kind of specific heat will in the following formulas be denoted by c for the first state of aggregation and by h for the second.[7] Then we have

$$\frac{dQ}{dT} = (M - m)c + mh.$$

We may also write this in the form

$$\frac{dQ}{dT} = m(h - c) + Mc. \tag{50}$$

From (49) and (50) it follows at once that

$$\frac{d}{dT}\left(\frac{dQ}{dm}\right) = \frac{dr}{dT}; \quad \frac{d}{dm}\left(\frac{dQ}{dT}\right) = h - c. \tag{51}$$

If we insert the foregoing expressions, Eqs. (48) to (51), into Eqs. (44), (45), and (46) after replacing m by x, we obtain

$$\frac{dr}{dT} + c - h = Au\,\frac{dp}{dT}, \tag{52}$$

$$\frac{dr}{dT} + c - h = \frac{r}{T}, \tag{53}$$

$$r = ATu\left(\frac{dp}{dT}\right). \tag{54}$$

[7]The letter c here therefore has a different meaning in the following formulas from that in the earlier ones, in which c denoted specific heat at constant volume.

The foregoing are the equations that, in my first article on the mechanical theory of heat, I derived as the principal equations related to the production of vapor.

In the numerical calculations that I made on the vaporization of water I did not distinguish the kind of specific heat involved in the above equations from the specific heat of water at constant pressure. This procedure is indeed fully justifiable, since in this case the difference between the two kinds of specific heat is smaller than the errors of observation![8]

[8] We can derive from the above equations the relation between the specific heat at constant pressure and that specific heat in which it is assumed that the pressure increases with temerature in such a way that it is always equal to the vapor pressure of the vapor produced from the liquid.

From Eq. (29) the quantity of heat that we must communicate to unit mass of liquid while the temperature changes by dT and the pressure changes by dp becomes

$$dQ = C\,dT - AT\left(\frac{dv}{dT}\right)_p dp,$$

where C denotes the specific heat at constant pressure. If we now imagine that the pressure increases with temperature like the maximum vapor pressure over the liquid and denote the pressure increase due to temperature rise dT by $(dp/dT)\,dT$, then the quantity of heat we must communicate to unit mass of the body under these circumstances in order to increase the temperature by dT is

$$dQ = C\,dT - AT\left(\frac{dv}{dT}\right)_p \cdot \left(\frac{dp}{dT}\right) dT\,.$$

If now we divide this equation by dT the resulting quotient dQ/dT is the specific heat in the sense meant here, denoted by c in the foregoing text. We then obtain

$$c = C - AT\left(\frac{dv}{dT}\right)_p \cdot \frac{dp}{dT}\,.$$

If we apply this to water and choose the temperature as 100°C, the thermal coefficient of expansion at 100°C, according to Kopp's investigation is 0.00080 if we take the volume of water at 4°C as unity. In order to obtain $(dv/dT)_p$ for the case in which the unit of volume is the cubic meter and the unit of mass is the kilogram, we must multiply the above figure by 0.001 and hence get

$$\left(\frac{dv}{dT}\right)_p = 0.00000080.$$

From the work of Regnault, if the pressure is given in kilograms per square meter at temperature 100°C, there results

$$\frac{dp}{dT} = 370.$$

For 100°C the absolute temperature is 373, and for A we take, following Joule, 1/424. Thus we obtain

If we form the total differential equation

$$dQ = \frac{dQ}{dm}\,dm + \frac{dQ}{dT}\,dT$$

$$AT\left(\frac{dv}{dT}\right)_p \cdot \frac{dp}{dT} = \frac{373}{424}\,(0.0000008)370 = 0.00026.$$

We note that the difference between C and c is too small for consideration in further calculations.

In considering the influence of pressure on the freezing point of liquids, the behavior is otherwise to the extent that a very considerable increase in pressure is required to change the freezing point only a little, and therefore the derivative dp/dT for this case has a very large value. The procedure that I followed in my first treatment of this matter (p. 33 of the first article in this collection) was somewhat inexact, since in my numerical calculations for this case I used the same values for c and h that people have hitherto used for water and ice at constant pressure. I must modify the remark I made in the appendix to that article that the difference between these quantities can be only very small. If we take the lowering of the freezing point for increase of pressure of one atmosphere as 0.00733°C we get

$$\frac{dp}{dT} = -\frac{10333}{0.00733}.$$

If we combine this value in the same fashion as previously with the thermal expansion coefficients of water and ice at 0°C, there result the following values in place of 1.0 and 0.48, which represent the normal values of the specific heat of water and ice respectively:

$$c = 1 - 0.05 = 0.95. \qquad h = 0.48 + 0.14 = 0.62.$$

If we insert these values in the equation

$$\frac{dr}{dT} = c - h + \frac{r}{T},$$

we get in place of our earlier value

$$\frac{dr}{dT} = 0.52 + 0.29 = 0.81$$

the following somewhat different result:

$$\frac{dr}{dT} = 0.33 + 0.29 = 0.62.$$

For the rest it should be remarked that this small correction applies only to an individual numerical calculation and has little practical significance. It is mentioned for theoretical interest only. The equation itself and the theoretical considerations connected with it are not essentially disturbed.

and introduce in it the expressions from (49) and (50) we get

$$dQ = r\,dm + [m\,(h - c) + Mc]\;dT\,.$$

If we insert here the expression for $h - c$ given in Eq. (53), the result is

$$dQ = r\,dm + \left[m\left(\frac{dr}{dT} - \frac{r}{T}\right) + Mc\right]\;dT\,,$$

which we can also write in the form

$$dQ = d\,(mr) - \frac{mr}{T}\,dT + Mc\,dT\,, \tag{55}$$

or even more concisely,

$$dQ = T\,d\left(\frac{mr}{T}\right) + Mc\,dT\,. \tag{56}$$

I shall not go into the applications of these equations here since they have been thoroughly discussed in my first articles and in the article on steam engines.

14. All the foregoing considerations refer to reversible changes. We now proceed to consider irreversible changes, at any rate to indicate briefly how they are to be handled.

In the mathematical investigations of irreversible processes we must consider two circumstances that give rise to characteristic magnitude determinations. In the first place, the quantities of heat that must be communicated to or taken away from a body in its change of state take place differently for irreversible processes than for reversible ones. In the second place, every irreversible change is connected with an uncompensated change, knowledge of which is of importance for certain considerations.

In order to provide appropriate analytic expressions for these two circumstances, it is necessary to recall some quantities in the equations set up previously.

One of these quantities, which is related to the first law, is the quantity U appearing in Eq. (1a) at the beginning of this article. This represents the heat and work content or the energy of the body. Equation (1a) is to be applied in the determination of this quantity. We can write this equation thus:

$$dU = dQ - dw\,, \tag{57}$$

or when integrated,

$$U = U_0 + Q - w\,. \tag{58}$$

In this equation U_0 represents the energy for an arbitrarily chosen initial state of the body, while Q and w mean respectively the quantity of heat that must be communicated to the body and the external work that is done while the body goes from that initial state to the final state through an arbitrary reversible process. As has already been said, the body (even when it is demanded that the change must be reversible) can proceed from the one state to the other in an infinite number of ways. Out of all these ways we are privileged to choose the one that is most convenient for calculation.

The other quantity to be considered here, relating to the second law, is contained in Eq. (2a). If indeed, as is expressed in this equation, the integral $\int(dQ/T)$ is always zero whenever the body performs a complete cycle of arbitrary processes back to its original state, the quantity under the integral sign, namely, dQ/T, must be the perfect differential of a quantity that depends only on the state the body happens to be in and not at all on the path by which it got there. If we denote this quantity by S, we may write

$$dS = \frac{dQ}{T}. \tag{59}$$

If we imagine the foregoing equation integrated for an arbitrary reversible process by which the body can go from the chosen initial state to its final state, and denote the value of S for the initial state by S_0, we have

$$S = S_0 + \int \frac{dQ}{T}. \tag{60}$$

This equation permits us to calculate S just as Eq. (58) allows us to determine U.

The physical meaning of the quantity S has been discussed in my article "On the Application of the Theorem of the Equivalence of Transformations to Internal Work of a Mass of Matter." The fundamental equation given in (II) of that article, which holds for all state changes of a body taking place reversibly, says that (if one makes the small alteration in its expression that one is treating as positive the heat *taken up* by the body rather than that given up by the body)

$$\int \frac{dQ}{T} = \int \frac{dH}{T} + \int dZ. \tag{61}$$

H and Z are new quantities introduced in the article just mentioned.

In the first integral on the right H is the heat actually present in the body, which, as I have shown, depends only on the temperature and not on the arrangement of the components of the body. From this it follows that the expression dH/T is an exact differential, and that when we form the integral $\int(dH/T)$ for the transition of the body from its initial to its final state, we obtain a quantity that is completely determined by these states, without needing to know the way in which the transition between the states has taken place. I have called this quantity (for reasons set forth in the above-mentioned article) the transformation value of the heat present in the body.

It would seem reasonable to choose the initial state of the body as that for which $H = 0$, that is, at the absolute zero of temperature. However, in this case the integral $\int(dH/T)$ becomes infinite. If we wish to obtain a finite value for the integral we must begin at our initial state for which the temperature has a nonvanishing value. The integral will not then represent the transformation value of all the heat contained in the body, but only the transformation value of the quantity of heat in the final state over and above that in the initial state. Thus I call the integral in question the transformation value of the body heat reckoned for the initial and final states. For brevity we shall denote this quantity as Y.

I have called the quantity Z in the second integral on the right of Eq. (61) the *disgregation* of the body. It depends on the arrangement of the components of the body. The measure of an increase in disgregation is the equivalence value of the transformation of work into heat that must take place in order to compensate for the increase in disgregation. It can accordingly serve as a replacement for the increase in disgregation. From this point of view we can say that the disgregation is the transformation value of the already existing arrangement of the components of the body. Since in the determination of the disgregation we must start from some initial state of the body, we shall assume that the initial state chosen is the same as that selected in the determination of the transformation value of the heat contained in the body.

If we now form the sum of the quantities Y and Z just defined, this is the quantity S as previously defined. If we go back to Eq. (61) and assume for the sake of generality that the initial state referred to by the integrals in this equation does not need to be precisely the same as that from which we started in the determination of Y and Z, but that here it is a question of a transformation whose initial state is wholly arbitrary, we can write for the integrals on the right-hand side of the equation

$$\int \frac{dH}{T} = Y - Y_0 \quad \text{and} \quad \int dZ = Z - Z_0,$$

in which Y_0 and Z_0 are the values of Y and Z that correspond to the initial state. Then Eq. (61) takes the form

$$\frac{dQ}{T} = Y + Z - (Y_0 - Z_0). \tag{62}$$

If in the foregoing we now set

$$Y + Z = S, \tag{63}$$

we get the equation

$$\int \frac{dQ}{T} = S - S_0, \tag{64}$$

which is essentially the same as Eq. (60), introduced for the determination of S.

We now seek an appropriate name for S. Just as we have called U the work content of the body, we would call S the transformation content of the body. However, I have felt it more suitable to take the names of important scientific quantities from the ancient languages in order that they may appear unchanged in all contemporary languages. Hence I propose that we call S the *entropy* of the body after the Greek word 'η τροπή, meaning "transformation." I have intentionally formed the word *entropy* to be as similar as possible to the word *energy*, since the two quantities that are given these names are so closely related in their physical significance that a certain likeness in their names has seemed appropriate.

For the sake of clarity let us stop to summarize the different quantities discussed in the course of this article. These are quantities that have been newly introduced by the mechanical theory of heat or have received by this theory an altered meaning. They have the common characteristic that they are determined by the momentarily existing state of the body without depending on our knowledge of the way in which the body arrived in this state. There are the following six quantities: (1) the heat content, (2) the work content, (3) the sum of the previous two quantities, called the energy of the body, (4) the transformation value of the heat content, (5) the disgregation, which is to be considered the existing arrangement of the components of the body, and (6) the sum of (4) and (5), accordingly the transformation content of the body or the entropy.

15. In order to calculate the energy and entropy for special cases, in addition to Eqs. (57) and (59) or (58) and (60), we have to use the different expressions for dQ given above. I shall here discuss only a few simple cases as examples.

Let us assume that the body under consideration is homogeneous and is at the same temperature throughout. We assume also that the only external force acting on it is a uniform and normal surface pressure, and that it can change its volume through a change in temperature and pressure without thereby suffering a change in its state of aggregation. If further the mass of the body is taken as a unit mass, for the determination of dQ we can apply Eqs. (28), (29), and (32) of paragraph 9. These equations contain the specific heat at constant volume c and the specific heat at constant pressure C. Since the latter is usually the one that is directly observed, we shall use the equation in which it occurs, namely Eq. (29), which has the form[9]

$$dQ = C\,dT - AT\,\frac{dv}{dT}\cdot dp.$$

So far as the external work is concerned, for an infinitely small change of state in which the volume changes by dv, we have

$$dw = Ap\,dv,$$

and since T and p have been chosen as the independent variables, we can put this equation in the form

$$dw = Ap\left(\frac{dv}{dT}\,dT + \frac{dv}{dp}\,dp\right).$$

If now we apply the foregoing expressions for dQ and dw to Eqs. (57) and (59), we obtain

[9] Instead of the form $(dv/dT)_p$ I am writing dv/dT simply in the above equation, since in a case in which T and p are the only independent variables it is understood that in differentiation with respect to T the other variable is constant.

$$dU = \left(C - AP \frac{dv}{dT}\right) dT - A\left(T \frac{dv}{dT} + p \frac{dv}{dp}\right) dp\,,$$
$$dS = \frac{C}{T}\, dT - A \frac{dv}{dT}\, dp\,. \tag{65}$$

Considering the second equation in (33), namely,

$$\frac{dC}{dp} = - AT \frac{d^2v}{dT^2}\,,$$

we readily convince ourselves that the two total differential equations in (65) are integrable, without the need for assuming a further relation between the variables. By carrying out the integration we obtain expressions for U and S in which each one contains an undetermined constant, namely, the values that the quantities U and S in question have at the initial state of the body chosen as the initial value for the integration.

If the body is an ideal gas the equations take a simple form. We can obtain them by combining Eqs. (65) with the equation $pv = RT$, expressing the content of the law of Mariotte and Gay-Lussac. Alternatively we can go back to Eqs. (57) and (59) and put in place of dQ one of the expressions already derived for ideal gases and contained in Eqs. (42), and at the same time insert for dw one of the three expressions: $AR\,(T/v)\,dv$; $AR\,[dT - (T/p)\,dp]$; $Ap\,dv$. If among the Eqs. (42) we choose the first, which is the simplest in the case in question, we get

$$dU = c\,dT\,,$$
$$dS = c\,\frac{dT}{T} + AR\,\frac{dv}{v}\,. \tag{66}$$

Since c and AR are constants we can at once carry out the integration of these equations. If we denote the initial values of U and S by U_0 and S_0 respectively (when $T = T_0$ and $v = v_0$), we obtain

$$U = U_0 + c\,(T - T_0),$$
$$S = S_0 + c \log \frac{T}{T_0} + AR \log \frac{v}{v_0}\,. \tag{67}$$

For the final special case we consider that to which paragraphs 12 and 13 relate, namely, where the body in question is a mass M of which the part $M - m$ is in one state of aggregation and the part m is in another, and in which the pressure on the whole body depends only on the temperature.

We shall assume that at the outset the whole M is in the first state of aggregation and is at temperature T_0 and under the pressure corresponding to this temperature. We denote the values of the energy and entropy in this state as U_0 and S_0 respectively. It is then supposed that the body is conveyed from this initial state to its final state by the following path. At first, while the whole mass stays in the first state of aggregation, the body is brought from the temperature T_0 to the temperature T and the pressure changes to the value corresponding to that temperature. Then at temperature T a part of the mass—namely, the part m—is assumed to go from the first state of aggregation to the second. We shall consider these two changes individually by applying the notation introduced in paragraph 13.

During the temperature change just mentioned we have to apply the equation

$$dQ = Mc\, dT\,.$$

The quantity c occurring here is the specific heat of the body in the first state of aggregation for the case in which the pressure varies during the temperature change in the manner indicated. We have discussed this kind of specific heat in the remark in paragraph 13 and, as has been pointed out there, we can use for the case in which the first state of aggregation is liquid or solid and the second state is vapor, the specific heat for liquid or solid bodies at constant pressure. It is only at very high temperatures at which the vapor pressure varies very rapidly with temperature that the difference between the specific heat c and that at constant pressure can become appreciable enough to need to take it into consideration. From the foregoing equation it follows that (if we recall that with the temperature increase dT a volume increase $M\,(d\sigma/dT)\,dT$ is associated and at the same time the external work becomes $MAp\,(d\sigma/dT)\,dT$

$$dU = M\left(c - Ap\,\frac{d\sigma}{dT}\right) dT\,,$$

$$dS = M\,\frac{c}{T}\,dT\,.$$

For the change in state of aggregation at temperature T we have

$$dQ = r\,dm.$$

Since the increase of the part in the second state of aggregation by dm brings about a volume increase of $u\,dm$ and at the same time external work represented by $Apu\,dm$, it follows that

$$dU = (r - Apu)\;dm.$$

In order to replace the quantity u with other quantities that are better known experimentally we apply Eq. (54) to the above, and obtain

$$Au = \frac{r}{T\,(dp/dT)}\,.$$

Hence

$$dU = r\left(1 - \frac{p}{T\,(dp/dT)}\right)\;dm.$$

At the same time from dQ we get at once for dS

$$dS = \frac{r}{T}\, dm.$$

Both differential equations relating to the first process must be integrated with respect to T from T_0 to T. Those relating to the second process must be integrated with respect to m from 0 to m. We then get

$$\begin{aligned} U &= U_0 + M \int_{T_0}^{T} \left(c - Ap \frac{d\sigma}{dT}\right) dT + mr\left(1 - \frac{p}{T\,(dp/dT)}\right), \\ S &= S_0 + M \int_{T_0}^{T} \frac{c}{T}\, dT + \frac{mr}{T}. \end{aligned} \tag{68}$$

16. If we now assume that the quantities U and S have been determined for a body in its different states by the method described above,[10] the equations that hold for irreversible processes can be written down at once.

The first fundamental equation (1a) and Eq. (58) obtained from this by integration that we may write in the form

$$Q = U - U_0 + w \tag{69}$$

hold just as well for irreversible processes as for reversible ones. The difference consists only in this: Of the quantities on the right-hand side of the equation the external work w has a different value in the case of an irreversible process than in the case of a reversible process leading to the same change. Such a difference does not exist for $U - U_0$, for this quantity is dependent only on the initial and final states and not on the path from the one state to the other. We need therefore to consider this path only so far as is necessary to determine the external work that is done. By adding the external work to $U - U_0$ we obtain the quantity of heat Q that the body must take up during the transition.

The uncompensated transformation associated with an irreversible change is handled in the following fashion.

The expression for the uncompensated transformation that can take place in a cyclic process has been given in Eq. (11) of my article[11] "On a Modified Form of the Second Law of the Mechanical Theory of Heat." In that article a quantity of heat *given up* by the body to a heat reservoir is considered positive, whereas in our present consideration heat *taken up* by the body to a heat reservoir is considered positive. We must now give dQ the sign opposite to that in the article mentioned. The equation in question becomes

[10] Some more complete mathematical developments on energy and entropy are communicated in an appendix to this article.

[11] *Pogg. Ann.* Vol. 93, p. 499. Also see Part One of the article collection, p. 115. [Not reproduced in this book.—*Ed.*]

$$N = - \int \frac{dQ}{T}. \qquad (70)$$

If now the body has suffered a change or a series of changes that do not form a cyclic process but in which a final state is reached that differs from the initial state, we can make a cyclic process out of this series of changes if we introduce additional changes of such a character that they enable the body to proceed from this final state back to the initial state. We assume that these additional changes are reversible.

If we apply Eq. (70) to the cyclic process so formed we can divide the integral into two parts, the first of which relates to the original transition from the initial to the final state and the second to the additional change taking the body back to its initial state. We shall write these two parts as separate integrals, distinguishing the second (corresponding to the transition back to the initial state) from the first by attaching the letter r to the integral sign. Thus Eq. (70) goes into the form

$$N = - \int \frac{dQ}{T} - \int_r \frac{dQ}{T}.$$

Since the change back to the initial state is assumed to be reversible, we can apply Eq. (64) to the second integral, with this difference only. If S_0 is the entropy of the initial state and S that of the final state, we must replace $S - S_0$ by $S_0 - S$, since here the process goes backward from the final to the initial state. We therefore write

$$\int \frac{dQ}{T} = S_0 - S.$$

With this substitution the preceding equation becomes

$$N = S - S_0 - \int \frac{dQ}{T}. \qquad (71)$$

The quantity N obtained in this way signifies the uncompensated transformation associated with the whole cyclic process. However, since for reversible processes we have the theorem that the sum of the transformations taking place in them is zero—that is, no uncompensated transformation can take place in them—the transition back to the initial state, which has been assumed reversible, can contribute nothing to the uncompensated transformation, and the quantity N therefore at once represents the uncompensated transformation of the body from the intiial to the final state. In the expression (71) the difference $S - S_0$ is completely determined if the initial and final states are given and it is only in the formation of the integral $\int (dQ/T)$ that the way in which the transition from the one state to the other takes place must be considered.

17. In conclusion I may be permitted to touch on a matter whose complete treatment is really not feasible here, since the necessary explanation would be too

extensive. However, I believe that the following brief sketch will not be without interest, since it can make clear the general importance of the quantities I have introduced in the formulation of the second principle of the mechanical theory of heat.

The second law in the form I have given it says that all transformations taking place in nature go by themselves in a certain direction, which I have denominated the positive sense. They can thus take place without compensation. They can take place in the opposite direction, that is, the negative, only when they are compensated at the same time by positive transformations. The application of this law to the universe as a whole leads to a conclusion to which W. Thomson first called attention[12] and which I have discussed in a recent article.[13] If in all changes of state taking place in the universe the transformations in one direction surpass in magnitude those taking place in the opposite direction, it follows that the total state of the universe will change continually in that direction and hence will inevitably approach a limiting state.

The question now arises how one can characterize this limiting state in simple yet definite fashion. We can do this in the following way: We can consider the transformations as mathematical quantities whose equivalence values can be calculated and joined together in a sum by algebraic addition.

I have carried out such calculations in my earlier articles with respect to the heat residing in bodies and the arrangement of their component parts. For every body there are available two quantities: the transformation value of its heat content and its disgregation, the sum of which forms its entropy. However, the matter is not completed there. We must also take into consideration radiant heat, or otherwise express the heat propagated through a space in the form of waves in the aether. Further, we must consider such motions as cannot be accommodated under the name *heat.*

The handling of the latter can be arranged, at least as far as the motions of ponderable masses are concerned, and leads through plausible considerations to the following conclusion. If a mass that is large enough so that an atom can be considered vanishingly small compared with it moves as a whole, the transformation value of its motion is to be considered vanishingly small compared with its kinetic energy. From this it follows that if such a motion is transformed into heat by the action of some passive resistance, the equivalence values of the uncompensated transformation associated therewith is simply represented by the transformation value of the heat developed. Radiant heat, however, cannot be handled in such a simple fashion. More particular considerations are needed to describe how its transformation value is to be calculated. Although in the earlier article just mentioned I discussed the connection between radiant heat and the mechanical theory of heat, I did not touch there on the question raised here, for in that earlier work I undertook to show only that there is no essential contradiction between the laws of radiant heat and a fundamental principle assumed by me in the mechanical theory of heat. The more special applications of the mechanical theory of heat and in

[12] *Phil. Mag.* 4th Series, Vol. 4, p. 304.
[13] *Pogg. Ann.* Vol. 121, p. 1 and Article 8 in my collection, part 1.

particular the application of the law of the equivalence of transformations to radiant heat I reserve for later treatment.

For now I shall restrict myself to the statement that if we imagine that the quantity that I have named the entropy of an individual body is formulated for the whole universe with due consideration of all attendant circumstances and if we join with this the somewhat simpler concept of energy, we can express the two fundamental principles of the mechanical theory of heat for the universe in the following simple form:

1. The energy of the universe is constant.
2. The entropy of the universe strives to attain a maximum value.

9

Reprinted from pp. 511–514 of *Mathematical and Physical Papers of William Thomson,* Vol. 1, Cambridge University Press, 1882, 571 pp.

ON A UNIVERSAL TENDENCY IN NATURE TO THE DISSIPATION OF MECHANICAL ENERGY*

W. Thomson

THE object of the present communication is to call attention to the remarkable consequences which follow from Carnot's proposition, that there is an absolute waste of mechanical energy available to man when heat is allowed to pass from one body to another at a lower temperature, by any means not fulfilling his criterion of a "perfect thermo-dynamic engine," established, on a new foundation, in the dynamical theory of heat. As it is most certain that Creative Power alone can either call into existence or annihilate mechanical energy, the "waste" referred to cannot be annihilation, but must be some transformation of energy†. To explain the nature of this transformation, it is convenient, in the first place, to divide *stores* of mechanical energy into two classes—*statical* and *dynamical.* A quantity of weights at a height, ready to descend and do work when wanted, an electrified body, a quantity of fuel, contain stores of mechanical energy of the statical kind. Masses of matter in motion, a volume of space through which undulations of light or radiant heat are passing, a body having thermal motions among its particles (that is, not infinitely cold), contain stores of mechanical energy of the dynamical kind.

The following propositions are laid down regarding the *dissipation* of mechanical energy from a given store, and the *restoration* of it to its primitive condition. They are necessary consequences of the axiom, "*It is impossible, by means of inanimate material agency, to derive mechanical effect from any portion of matter by cooling it below the temperature of the coldest of the surrounding objects.*" (Dynamical Theory of Heat [Art. XLVIII. above], § 12.)

* From the Proceedings of the Royal Society of Edinburgh for April 19, 1852, also *Philosophical Magazine*, Oct. 1852.

† See the Author's previous paper on the Dynamical Theory of Heat, § 22 [Art. XLVIII. above].

I. When heat is created by a reversible process (so that the mechanical energy thus spent may be *restored* to its primitive condition), there is also a transference from a cold body to a hot body of a quantity of heat bearing to the quantity created a definite proportion depending on the temperatures of the two bodies.

II. When heat is created by any unreversible process (such as friction), there is a *dissipation* of mechanical energy, and a full *restoration* of it to its primitive condition is impossible.

III. When heat is diffused by *conduction*, there is a *dissipation* of mechanical energy, and perfect *restoration* is impossible.

IV. When radiant heat or light is absorbed, otherwise than in vegetation, or in chemical action, there is a *dissipation* of mechanical energy, and perfect *restoration* is impossible.

In connexion with the second proposition, the question, *How far is the loss of power experienced by steam in rushing through narrow steam-pipes compensated, as regards the economy of the engine, by the heat* (containing an exact equivalent of mechanical energy) *created by the friction?* is considered, and the following conclusion is arrived at:—

Let S denote the temperature of the steam (which is nearly the same in the boiler and steam-pipe, and in the cylinder till the expansion within it commences); T the temperature of the condenser; μ the value of Carnot's function, for any temperature t; and R the value of

$$\epsilon^{-\frac{1}{J}\int_T^S \mu dt}.$$

Then $(1-R)w$ expresses the greatest amount of mechanical effect that can be economized in the circumstances from a quantity w/J of heat produced by the expenditure of a quantity w of work in friction, whether of the steam in the pipes and entrance ports, or of any solids or fluids in motion in any part of the engine; and the remainder, Rw, is absolutely and irrecoverably wasted, unless some use is made of the heat discharged from the condenser. The value of $1-R$ has been shown to be not more than about $\frac{1}{4}$ for the best steam-engines, and we may infer that in them at least three-fourths of the work spent in any kind of friction is utterly wasted.

In connexion with the third proposition, the quantity of work that could be got by equalizing the temperature of all parts of a solid body possessing initially a given non-uniform distribution of heat, if this could be done by means of perfect thermo-dynamic engines without any conduction of heat, is investigated. If t be the initial temperature (estimated according to any arbitrary system) at any point xyz of the solid, T the final uniform temperature, and c the thermal capacity of unity of volume of the solid the required mechanical effect is of course equal to

$$J\iiint c(t-T)\,dx\,dy\,dz,$$

being simply the mechanical equivalent of the amount of heat put out of existence. Hence the problem becomes reduced to that of the determination of T. The following solution is obtained:—

$$T=\frac{\iiint \epsilon^{-\frac{1}{J}\int_0^t \mu dt}\,ct\,dx\,dy\,dz}{\iiint \epsilon^{-\frac{1}{J}\int_0^t \mu dt}\,c\,dx\,dy\,dz}.$$

If the system of thermometry adopted* be such that $\mu=\dfrac{J}{t+a}$, that is, if we agree to call $J/\mu-a$ the *temperature* of a body, for which μ is the *value of Carnot's function* (a and J being constants), the preceding expression becomes

$$T=\frac{\iiint c\,dx\,dy\,dz}{\iiint \dfrac{c}{t+a}\,dx\,dy\,dz}-a.$$

The following general conclusions are drawn from the propositions stated above, and known facts with reference to the mechanics of animal and vegetable bodies:—

* According to "Mayer's hypothesis," this system coincides with that in which equal differences of temperature are defined as those with which the same mass of air under constant pressure has equal differences of volume, provided J be the mechanical equivalent of the thermal unit, and a^{-1} the coefficient of expansion of air.

1. There is at present in the material world a universal tendency to the dissipation of mechanical energy.

2. Any *restoration* of mechanical energy, without more than an equivalent of dissipation, is impossible in inanimate material processes, and is probably never effected by means of organized matter, either endowed with vegetable life or subjected to the will of an animated creature.

3. Within a finite period of time past, the earth must have been, and within a finite period of time to come the earth must again be, unfit for the habitation of man as at present constituted, unless operations have been, or are to be performed, which are impossible under the laws to which the known operations going on at present in the material world are subject.

10

Reprinted from pp. 515–520 of *Mathematical and Physical Papers of William Thomson,* Vol. 1, Cambridge University Press, 1882, 571 pp.

ON THE ECONOMY OF THE HEATING OR COOLING OF BUILDINGS BY MEANS OF CURRENTS OF AIR*

W. Thomson

IF it be required to introduce a certain quantity of air at a stated temperature higher than that of the atmosphere into a building, it might at first sight appear that the utmost economy would be attained if all the heat produced by the combustion of the coals used were communicated to the air; and in fact the greatest economy that has yet been aimed at in heating air or any other substance, for any purpose whatever, has had this for its limit. If an engine be employed to pump in air for heating and ventilating a building (as is done in Queen's College, Belfast), all the waste heat of the engine, along with the heat of the fire not used in the engine, may be applied by suitable arrangements to warm the entering current of air; and even the heat actually converted into mechanical effect by the engine, will be reconverted into heat by the friction of the air in the passages, since the overcoming of resistance depending on this friction is the sole work done by the engine. It appears therefore that whether the engine be economical as a converter of heat into mechanical work, or not, there would be perfect economy of the heat of the fire if all the heat escaping in any way from the engine, as well as all the residue from the fire, were applied to heating the air pumped in, and if none of this heat were

* Mathematical demonstrations of the results stated in this paper have since been published in the *Camb. and Dub. Math. Journal,* Nov. 1853. [Art. XLVIII. note III. above.]

allowed to escape by conduction through the air passages. It is not my present object to determine how nearly in practice this degree of economy may be approximated to; but to point out how the limit which has hitherto appeared absolute, may be surpassed, and a current of warm air at such a temperature as is convenient for heating and ventilating a building may be obtained mechanically, either by water power without any consumption of coals, or, by means of a steam engine, driven by a fire burning actually less coals than are capable of generating by their combustion the required heat; and secondly, to show how, with similar mechanical means, currents of cold air, such as might undoubtedly be used with great advantage to health and comfort for cooling houses in tropical countries*, may be produced by motive power requiring (if derived from heat by means of steam engines) the consumption of less coals perhaps than are used constantly for warming houses in this country.

In the mathematical investigation communicated with this paper, it is shown in the first place, according to the general principles of the dynamical theory of heat, that any substance may be heated thirty degrees above the atmospheric temperature by means of a properly contrived machine, driven by an agent spending not more than about $\frac{1}{35}$ of energy of the heat thus communicated; and that a corresponding machine, or the same machine worked backwards, may be employed to produce cooling effects, requiring about the same expenditure of energy in working it to cool the same substance through a similar range of temperature. When a body is heated by such means, about $\frac{34}{35}$

* The mode of action and apparatus proposed for this purpose differs from that proposed originally by Professor Piazzi Smyth for the same purpose, only in the use of an egress cylinder, by which the air is made to do work by its extra pressure and by expansion in passing from the reservoir to the locality where it is wanted, which not only saves a great proportion of the motive power that would be required were the air allowed simply to escape through a passage, regulated by a stop-cock or otherwise, but is absolutely essential to the success of the project, as it has been demonstrated by Mr Joule and the author of this communication, that the cold of expansion would be so nearly compensated by the heat generated by friction, when the air is allowed to rush out without doing work, as to give not two-tenths of a degree of cooling effect in apparatus planned for 30 degrees. The use of an egress cylinder has (as the meeting was informed by Mr Macquorn Rankine), recently been introduced into plans adopted by a committee of the British Association appointed to consider the practicability of Professor Piazzi Smyth's suggestion, with a view to recommending it to government for public buildings in India.

of the heat is drawn from surrounding objects, and $\frac{1}{35}$ is created by the action of the agent; and when a body is cooled by the corresponding process, the whole heat abstracted from it, together with a quantity created by the agent, equal to about $\frac{1}{35}$ of this amount, is given out to the surrounding objects.

A very good steam engine converts about $\frac{1}{10}$ of the heat generated in its furnace into mechanical effect; and consequently, if employed to work a machine of the kind described, might raise a substance thirty degrees above the atmospheric temperature by the expenditure of only $\frac{10}{35}$, or $\frac{2}{7}$, that is, less than one-third of the coal that would be required to produce the same elevation of temperature with perfect economy in a direct process. If a water-wheel were employed, it would produce by means of the proposed machine the stated elevation of temperature, with the expenditure of $\frac{1}{35}$ of the work, which it would have to spend to produce the same heating effect by friction.

The machine by which such effects are to be produced must have the properties of a "perfect thermo-dynamic engine," and in practice would be either like a steam engine, founded on the evaporation and re-condensation of a liquid (perhaps some liquid of which the boiling point is lower than that of water), or an air engine of some kind. If the substance to be heated or cooled be air, it will be convenient to choose this itself as the medium operated on in the machine. For carrying out the proposed object, including the discharge of the air into the locality where it is wanted, the following general plan was given as likely to be found practicable. Two cylinders, each provided with a piston, ports, valves, and expansion gearing, like a high-pressure double-acting steam engine, are used, one of them to pass air from the atmosphere into a large receiver, and the other to remove air from this receiver and discharge into the locality where it is wanted. The first, or ingress cylinder and the receiver, should be kept with their contents as nearly as possible at the atmospheric temperature, and for this purpose ought to be of good conducting material, as thin as is consistent with the requisite strength, and formed so as to expose as much external surface as possible to the atmosphere, or still better, to a stream of water. The egress cylinder ought to be protected as much as possible from thermal communication with the atmosphere or surrounding objects. According as the air is to be heated, or cooled, the pistons and valve gearing must be worked so as to keep the pressure in the receiver below, or above, that of the atmosphere. If the cylinders be of equal dimensions, the arrangement when the air is to be heated, would be as follows:—The two pistons working at the same rate, air is to be admitted freely from the atmosphere into the ingress cylinder, until a certain fraction of the stroke, depending on the heating effect required, is performed, then the entrance port is to be shut, so that during the remainder of the stroke the air may expand down to the pressure of the receiver, into which, by the opening of another valve, it is to be admitted in the reverse stroke; while the egress cylinder* is to draw air freely from the receiver through the whole of each stroke on one side or the other of its piston, and in the reverse stroke first to compress this air to the atmospheric pressure (and so heat it as required), and then discharge it into a pipe leading to the locality where it is to be used. If it be required to heat the air from 50° to 80° Fahr., the ratio of expansion to the whole stroke in the ingress cylinder would be $\frac{18}{100}$, the pressure of the air in the receiver would be $\frac{82}{100}$ of that of the atmosphere (about 2·7 lbs. on the square inch below the atmospheric pressure), and the ratio of compression to the whole stroke in the egress cylinder would be $\frac{13}{100}$. If 1 lb. of air (or about $13\frac{1}{2}$ cubic feet, at the stated temperature of 80°, and the mean atmospheric pressure,) be to be delivered per second, the motive power required for working the machine would be ·283 of a horse power, were the action perfect, with no loss of effect, by friction, by loss of expansive power due to cooling in the ingress cylinder, or otherwise. If each cylinder be four feet in stroke, and 26·3 inches diameter, the pistons would have to be worked at 26·1 double strokes per minute.

On the other hand, if it be desired to cool air, either the ingress piston must be worked faster than the other, or the stroke

* In this case the egress cylinder acts merely as an air pump, to draw air from the receiver and discharge it into the locality where it is wanted, and the valves required for this purpose might be ordinary self-acting pump-valves. A similar remark applies to the action of the ingress cylinder in the use of the apparatus for producing a cooling effect on the air transmitted, which will then be that of a compressing air-pump to force air from the atmosphere into the receiver. But in order that the same apparatus may be used for the double purpose of heating or cooling, as may be required at different seasons, it will be convenient to have the valves of each cylinder worked mechanically, like those of a steam engine.

of the other must be diminished, or the ingress cylinder must be larger, or an auxiliary ingress cylinder must be added. The last plan appears to be undoubtedly the best, as it will allow the two principal pistons to be worked stroke for stroke together, and consequently to be carried by one piston rod, or by a simple lever, without the necessity of any variable connecting gearing, whether the machine be used for heating or for cooling air; all that is necessary to adapt it to the latter purpose, besides altering the valve gearing, being to connect a small auxiliary piston to work beside the principal ingress cylinder, with which it is to have free communication at each end. If it were required to cool air from 80° to 50° Fahr., the auxiliary cylinder would be required to have its volume $\frac{1}{17}$ of that of each of the principal cylinders; and, if its stroke be the same, its diameter would therefore be a little less than a quarter of theirs. The valves would have to be altered to give compression in the ingress cylinder during the same fraction of the stroke as is required for expansion when the air is heated through the same range of temperature, and the valves of the egress cylinder would have to give the same proportion of expansion as is given of compression in the other case; and the pressure kept up in the receiver, by the action of the pistons thus arranged, would be $1\frac{18}{100}$ atmos., or about 3·2 lbs. on the square inch above the atmospheric pressure. The principal cylinders being of the same dimensions as those assumed above, and the quantity of air required being the same (1 lb. per second), the pistons would have to be worked at only 21·4 double strokes per minute instead of 26·1, and the horse power required would be ·288, instead of as formerly ·283, when the same machine was used for giving a supply of heated air.

[Note added June 26, 1881. The method of cooling air in unlimited quantities described in this article has been realized by Mr Coleman, first in refrigerators used for the distillation of paraffin, and after that in the Bell-Coleman refrigerator, for carrying supplies of fresh meat from North America to Europe; in a great refrigerator recently sent out for the Abattoir at Brisbane, Queensland; and other large practical applications of a similar kind. The Bell-Coleman machine sends large quantities of air cooled 10° or 20° C. below freezing point into the chamber to be kept cool; and the general temperature of this chamber is thus maintained at the desired point, which, for the case of carrying fresh meat from America to this country is about 35° F.

The method of heating air described in the article remains unrealized to this day. When Niagara is set to work for the benefit of North America through electric conductors, it will no doubt be largely employed for the warming of houses over a considerable part of Canada and the United States. But it is probable that it will also have applications though less large in other cold countries, to multiply the heat of coal and other fuel, and to utilize wind and water power (with aid of electric accumulators) for warming houses.]

Part III

THE ROUNDING-OFF

Editor's Comments on Papers 11 Through 14

11 **GIBBS**
On the Equilibrium of Heterogeneous Substances

12 **CARATHÉODORY**
Investigation into the Foundations of Thermodynamics

13 **PLANCK**
Second Law of Thermodynamics

14 **BORN**
Antecedence: Thermodynamics

The first two papers of Part III of this volume complete the development of the second law as we know it today. They can stand on their own feet because except for the notation, they are written in a thoroughly modern style.

Paper 11 by Gibbs is his own abstract of the magnum opus of the same title that appeared between 1875 and 1878 in the *Transactions of the Connecticut Academy*, occupying about three hundred pages in it. The paper reproduced here contains a terse statement of the equilibrium principle and some of its applications. The major work (Gibbs 1961) is easily accessible, constitutes a reference book in itself, and can be—indeed ought to be—studied by everybody who is interested in thermodynamics.

The extension of our knowledge achieved by Carathéodory in Paper 12 is of a different nature. Many contemporary writers, like some at the beginning of this century, would characterize the fundamental papers reproduced so far as lacking in rigor—meaning, in effect, lacking in mathematical rigor and generality. For example, practically all fundamental papers operate with pure substances that are characterized by two independent variables and by the single term $p\,dv$ in the expression for work in a reversible process. Now it is known from the theory of Pfaff's differential equations that *every* linear differential form in *two* independent variables possesses an integrating factor. Thus in this light, the early development, if it were truly restricted to such simple systems, would appear like an attempt to crash through an open door when the existence of entropy is established. The only solid achievement remaining would be the assertion that the integrating denominator for

$$dQ^\circ = du + p\,dv \tag{1}$$

(which is sure to exist) can be normalized to become a unique function $T(t)$ of any arbitrary empirical temperature t.

The early writers seem to have had no doubts whatever that the second law applied to the most complex systems imaginable, but the question remained: Could this fact be laid bare in a more convincing fashion than by means of an intuitive extension of its validity into territory explored only inadequately?

The required reasoning, and rigor, have been supplied by the mathematician Carathéodory on the prompting of the physicist Born, as will be revealed later (Paper 14).

Carathéodory's starting point was a development of Max Planck's exposition of the second law, which I have included as Paper 13. Carathéodory succeeded in extracting the quintessential physical content of the basis for the second law, which in plain words amounts to saying that all real processes are irreversible. He expressed this as the following axiom:

> *In every arbitrarily close neighborhood of a given initial state there exist states which cannot be approached arbitrarily closely by adiabatic processes.*

There are no restrictions regarding the nature of the system, and it is recognized from the outset that the element of work in a reversible process is given by

$$dQ^\circ = d\epsilon + DA \tag{1a}$$

where ϵ denotes energy, and where

$$DA = p_1\,dx_1 + p_2\,dx_2 + \ldots + p_n\,dx_n; \tag{2}$$

here the x_i's denote *deformation variables*; the only thing still missing, compared with our contemporary views, is a clear distinction (rather than implied, as above) between internal and external deformation variables.

The great achievement that we owe to Carathéodory is the very creative observation of the existence of a link between two sets of seemingly disparate conceptions. First, he noticed the link between the existence of an integrating denominator of a linear differential form and of the mathematical idea of accessibility in relation to Pfaff's equation

$$d\epsilon + DA = 0. \tag{3}$$

This yields Carathéodory's mathematical theorem.[3] Second, he noticed the link between irreversibility and physical accessibility of thermodynamic states and thus was led to rephrase the second law and to give it the form of the preceding axiom.

Since now the thermodynamic system under consideration can possess an arbitrary (though finite) number of degrees of freedom, the assertion that *any* such system is of a type to admit an integrating denominator for its $dQ°$ ceases to be trivial and acquires fundamental importance. The task of showing that the thermodynamic temperature can be chosen as this integrating denominator still remains, and this Carathéodory modeled on Planck.

I have included Chapter III, "Second Law of Thermodynamics," of Planck's *Theory of Heat* for several reasons. First, its clarity is admirable, and the chapter is worth reading for its own sake. Even though the text has been taken from the 1932 edition, it is necessary to realize that the first formulation of his ideas on this subject occurred before 1897, which is when he worked on the first edition of his *Vorlesungen über Thermodynamik.* It is most likely that Carathéodory knew this book, but I prefer to quote Planck's version of the second law somewhat later rather than earlier, to bring into clearer relief the principal line of development from Kelvin-Clausius, through Gibbs to Carathéodory.

Planck stated the second law in a manner reminiscent of Lord Kelvin:

> *It is impossible to construct a machine which functions with a regular period and which does nothing but raise a weight and causes a corresponding cooling of a heat reservoir.*

He then proceeded to consider two states Z and Z' of an arbitrary system of bodies and to examine processes that transfer the system from the state Z to the state Z', or conversely, "without anything being left changed outside the system," that is, adiabatically. In such circumstances it is possible to show that only one of the two processes $Z \rightarrow Z'$ or $Z' \rightarrow Z$ is possible unless Z and Z' are specially chosen. In the former case the process is irreversible; it is reversible in the latter. In other words, one of the two states is inaccessible from the other by

[3] A detailed and rather elementary proof can be found in J. Kestin, *A Course in Thermodynamics*, vol. I, Blaisdell, Waltham, Mass., 1965. See Chapter 10, pp. 457–503.

means of an adiabatic process. Carathéodory took this as his enunciation of the second law, as we have seen.

Carathéodory's proof that the integrating denominator for (1a)—whose existence he so brilliantly demonstrated—can be chosen to be a unique function of any empirical temperature τ is equivalent to Planck's given on p. 61 (**268**) and believed to be the first.

It appears that Carathéodory's work did not meet with early recognition. For example, Sommerfeld in his *Thermodynamics and Statistical Mechanics* (1956) calls it "even more abstract [than Planck's] and, at the same time, simpler if the simplicity of a proof is judged by the small number of assumptions required." He then approvingly repeats Carathéodory's own doubts, which Carathéodory expressed in his inaugural address to the Prussian Academy in 1919:

> The resulting theory is logically unassailable and satisfactory for the mathematician . . . . And yet, precisely these merits impede its usefulness to the student of nature, because, on the one hand, temperature appears as a derived quantity, and . . . it is impossible to establish a connection between the world of visible and tangible matter and the world of atoms through the smooth walls of the all too artificial structure.

Carathéodory's work was sparked by Born, as Born himself says on p. 38 (**298**): "I discussed the problem with my mathematical friend, Carathéodory, with the result that he analyzed it and produced a much more satisfactory solution" [to the lack of mathematical rigor at the foundations of thermodynamics as taught at the time]. Nevertheless, Born felt it necessary to explain the method in clearer terms. Born (1921) did this in his article in *Physikalische Zeitschrift*, whose contents he later paraphrased when he delivered in 1948 his Waynflete Lectures at St. Mary Magdalen at Oxford University. By general consent, and in Carathéodory's own judgment, the lecture reproduced here as Paper 14, together with the associated appendixes, constitutes one of the most lucid presentations of the second law by the elegant method of Carathéodory.

At this point it is possible to assert that the development of the second law, as it exists in its classical form, has been satisfactorily concluded. However, search as we may through the seminal papers produced so far, we must conclude that no serious effort has been devoted to a systematic study of the applicability of the second law to *irreversible processes* conceived as continuous sequences of nonequilibrium states. Born points to this deficiency clearly in Appendix 10, given on p. 151 (**310**). Thus our Part III ends with the recognition that a new terrain needs to be charted: the theory of irreversible processes as they occur, necessarily, in continuous systems when their states are removed from thermodynamic equilibrium. This quest is still in progress, and we shall document its status in a separate volume.

REFERENCES

Born, M. 1921. Kritische Betrachtungen zur traditionellen Darstellung der Thermodynamik. *Phys. Zeitschr.* **22**: 218–224, 249–254, 282–285.

Carathéodory, C. 1919. Antrittsrede. *Sitzber. Preuss. Akad. Wiss.* Art. XXXIII: 566–568.

Gibbs, J. W. 1961. *The Scientific Papers of J. Willard Gibbs, Vol. I.* Dover, New York, pp. 55–353.

Sommerfeld, A. 1956. *Lectures in Theoretical Physics, Vol. V: Thermodynamics and Statistical Mechanics*, J. Kestin, trans., Academic Press, New York, p. 39.

11

Reprinted by permission of the publisher from pp. 354–371 of *The Scientific Papers of J. W. Gibbs, Vol. I: Thermodynamics*, Dover Publications, Inc., New York, 1961, 462 pp.

ON THE EQUILIBRIUM OF HETEROGENEOUS SUBSTANCES.

J. W. Gibbs

Abstract of the Preceding Paper by the Author.

[*American Journal of Science*, 3 ser., vol. XVI., pp. 441–458, Dec., 1878.]

It is an inference naturally suggested by the general increase of entropy which accompanies the changes occurring in any isolated material system that when the entropy of the system has reached a maximum, the system will be in a state of equilibrium. Although this principle has by no means escaped the attention of physicists, its importance does not appear to have been duly appreciated. Little has been done to develop the principle as a foundation for the general theory of thermodynamic equilibrium.

The principle may be formulated as follows, constituting a criterion of equilibrium :—

I. *For the equilibrium of any isolated system it is necessary and sufficient that in all possible variations of the state of the system which do not alter its energy, the variation of its entropy shall either vanish or be negative.*

The following form, which is easily shown to be equivalent to the preceding, is often more convenient in application :—

II. *For the equilibrium of any isolated system it is necessary and sufficient that in all possible variations of the state of the system which do not alter its entropy, the variation of its energy shall either vanish or be positive.*

If we denote the energy and entropy of the system by ϵ and η respectively, the criterion of equilibrium may be expressed by either of the formulæ

$$(\delta\eta)_\epsilon \leqq 0, \tag{1}$$

$$(\delta\epsilon)_\eta \geqq 0. \tag{2}$$

Again, if we assume that the temperature of the system is uniform, and denote its absolute temperature by t, and set

$$\psi = \epsilon - t\eta, \tag{3}$$

the remaining conditions of equilibrium may be expressed by the formula

$$(\delta\psi)_t \geqq 0, \tag{4}$$

the suffixed letter, as in the preceding cases, indicating that the quantity which it represents is constant. This condition, in connection with that of uniform temperature, may be shown to be equivalent to (1) or (2). The difference of the values of ψ for two different states of the system which have the same temperature represents the work which would be expended in bringing the system from one state to the other by a reversible process and without change of temperature.

If the system is incapable of thermal changes, like the systems considered in theoretical mechanics, we may regard the entropy as having the constant value zero. Conditions (2) and (4) may then be written

$$\delta\epsilon \geqq 0, \qquad \delta\psi \geqq 0,$$

and are obviously identical in signification, since in this case $\psi = \epsilon$.

Conditions (2) and (4), as criteria of equilibrium, may therefore both be regarded as extensions of the criterion employed in ordinary statics to the more general case of a thermodynamic system. In fact, each of the quantities $-\epsilon$ and $-\psi$ (relating to a system without sensible motion) may be regarded as a kind of force-function for the system,—the former as the force-function *for constant entropy* (i.e., when only such states of the system are considered as have the same entropy), and the latter as the force-function *for constant temperature* (i.e., when only such states of the system are considered as have the same uniform temperature).

In the deduction of the particular conditions of equilibrium for any system, the general formula (4) has an evident advantage over (1) or (2) with respect to the brevity of the processes of reduction, since the limitation of constant temperature applies to every part of the system taken separately, and diminishes by one the number of independent variations in the state of these parts which we have to consider. Moreover, the transition from the systems considered in ordinary mechanics to thermodynamic systems is most naturally made by this formula, since it has always been customary to apply the principles of theoretical mechanics to real systems on the supposition (more or less distinctly conceived and expressed) that the temperature of the system remains constant, the mechanical properties of a thermodynamic system maintained at a constant temperature being such as might be imagined to belong to a purely mechanical system, and admitting of representation by a force-function, as follows directly from the fundamental laws of thermodynamics.

Notwithstanding these considerations, the author has preferred in general to use condition (2) as the criterion of equilibrium, believing that it would be useful to exhibit the conditions of equilibrium of thermodynamic systems in connection with those quantities which are most simple and most general in their definitions, and which appear most important in the general theory of such systems. The slightly different form in which the subject would develop itself, if condition (4) had been chosen as a point of departure instead of (2), is occasionally indicated.

Equilibrium of masses in contact.—The first problem to which the criterion is applied is the determination of the conditions of equilibrium for different masses in contact, when uninfluenced by gravity, electricity, distortion of the solid masses, or capillary tensions. The statement of the result is facilitated by the following definition.

If to any homogeneous mass in a state of hydrostatic stress we suppose an infinitesimal quantity of any substance to be added, the mass remaining homogeneous and its entropy and volume remaining unchanged, the increase of the energy of the mass divided by the quantity of the substance added is the *potential* for that substance in the mass considered.

In addition to equality of temperature and pressure in the masses in contact, it is necessary for equilibrium that the potential for every substance which is an independently variable component of any of the different masses shall have the same value in all of which it is such a component, so far as they are in contact with one another. But if a substance, without being an actual component of a certain mass in the given state of the system, is capable of being absorbed by it, it is sufficient if the value of the potential for that substance in that mass is not less than in any contiguous mass of which the substance is an actual component. We may regard these conditions as sufficient for equilibrium with respect to infinitesimal variations in the composition and thermodynamic state of the different masses in contact. There are certain other conditions which relate to the possible formation of masses entirely different in composition or state from any initially existing. These conditions are best regarded as determining the stability of the system, and will be mentioned under that head.

Anything which restricts the free movement of the component substances, or of the masses as such, may diminish the number of conditions which are necessary for equilibrium.

Equilibrium of osmotic forces.—If we suppose two fluid masses to be separated by a diaphragm which is permeable to some of the component substances and not to others, of the conditions of equilibrium which have just been mentioned, those will still subsist which relate to temperature and the potentials for the substances to which the diaphragm is permeable, but those relating to the potentials for the substances to which the diaphragm is impermeable will no longer be necessary. Whether the pressure must be the same in the two fluids will depend upon the rigidity of the diaphragm. Even when the diaphragm is permeable to all the components without restriction, equality of pressure in the two fluids is not always necessary for equilibrium.

Effect of gravity.—In a system subject to the action of gravity, the potential for each substance, instead of having a uniform value throughout the system, so far as the substance actually occurs as an independently variable component, will decrease uniformly with increasing height, the difference of its values at different levels being equal to the difference of level multiplied by the force of gravity.

Fundamental equations.—Let ϵ, η, v, t and p denote respectively the energy, entropy, volume, (absolute) temperature, and pressure of a homogeneous mass, which may be either fluid or solid, provided that it is subject only to hydrostatic pressures, and let $m_1, m_2, \ldots m_n$ denote the quantities of its independently variable components, and $\mu_1, \mu_2, \ldots \mu_n$ the potentials for these components. It is easily shown that ϵ is a function of $\eta, v, m_1, m_2, \ldots m_n$, and that the complete value of $d\epsilon$ is given by the equation

$$d\epsilon = t\,d\eta - p\,dv + \mu_1 dm_1 + \mu_2 dm_2 \ldots + \mu_n dm_n. \tag{5}$$

Now if ϵ is known in terms of $\eta, v, m_1, \ldots m_n$, we can obtain by differentiation $t, p, \mu_1, \ldots \mu_n$ in terms of the same variables. This will make $n+3$ independent known relations between the $2n+5$ variables, $\epsilon, \eta, v, m_1, m_2, \ldots m_n, t, p, \mu_1, \mu_2, \ldots \mu_n$. These are all that exist, for of these variables, $n+2$ are evidently independent. Now upon these relations depend a very large class of the properties of the compound considered,—we may say in general, all its thermal, mechanical, and chemical properties, so far as *active tendencies* are concerned, in cases in which the form of the mass does not require consideration. A single equation from which all these relations may be deduced may be called a fundamental equation. An equation between $\epsilon, \eta, v, m_1, m_2, \ldots m_n$ is a fundamental equation. But there are other equations which possess the same property.

If we suppose the quantity ψ to be determined for such a mass as we are considering by equation (3), we may obtain by differentiation and comparison with (5)

$$d\psi = -\eta\,dt - p\,dv + \mu_1 dm_1 + \mu_2 dm_2 \ldots + \mu_n dm_n. \tag{6}$$

If, then, ψ is known as a function of $t, v, m_1, m_2, \ldots m_n$, we can find $\eta, p, \mu_1, \mu_2, \ldots \mu_n$ in terms of the same variables. If we then substitute for ψ in our original equation its value taken from equation (3) we shall have again $n+3$ independent relations between the same $2n+5$ variables as before.

Let

$$\zeta = \epsilon - t\eta + pv, \tag{7}$$

then, by (5),

$$d\zeta = -\eta\, dt + v\, dp + \mu_1 dm_1 + \mu_2 dm_2 \ldots + \mu_n dm_n. \tag{8}$$

If, then, ζ is known as a function of $t, p, m_1, m_2, \ldots m_n$, we can find $\eta, v, \mu_1, \mu_2, \ldots \mu_n$ in terms of the same variables. By eliminating ζ, we may obtain again $n+3$ independent relations between the same $2n+5$ variables as at first.*

If we integrate (5), (6) and (8), supposing the quantity of the compound substance considered to vary from zero to any finite value, its nature and state remaining unchanged, we obtain

$$\epsilon = t\eta - pv + \mu_1 m_1 + \mu_2 m_2 \ldots + \mu_n m_n, \tag{9}$$

$$\psi = -pv + \mu_1 m_1 + \mu_2 m_2 \ldots + \mu_n m_n, \tag{10}$$

$$\zeta = \mu_1 m_1 + \mu_2 m_2 \ldots + \mu_n m_n. \tag{11}$$

If we differentiate (9) in the most general manner, and compare the result with (5), we obtain

$$-v\, dp + \eta\, dt + m_1 d\mu_1 + m_2 d\mu_2 \ldots + m_n d\mu_n = 0, \tag{12}$$

or

$$dp = \frac{\eta}{v} dt + \frac{m_1}{v} d\mu_1 + \frac{m_2}{v} d\mu_2 \ldots + \frac{m_n}{v} d\mu_n = 0. \tag{13}$$

Hence, there is a relation between the $n+2$ quantities $t, p, \mu_1, \mu_2, \ldots \mu_n$, which, if known, will enable us to find in terms of these quantities all the ratios of the $n+2$ quantities $\eta, v, m_1, m_2, \ldots m_n$. With (9), this will make $n+3$ independent relations between the same $2n+5$ variables as at first.

Any equation, therefore, between the quantities

	ϵ,	η,	v,	m_1,	$m_2, \ldots m_n$,
or	ψ,	t,	v,	m_1,	$m_2, \ldots m_n$,
or	ζ,	t,	p,	m_1,	$m_2, \ldots m_n$,
or		t,	p,	μ_1,	$\mu_2, \ldots \mu_n$,

is a fundamental equation, and any such is entirely equivalent to any other.

Coexistent phases.—In considering the different homogeneous bodies which can be formed out of any set of component substances, it is convenient to have a term which shall refer solely to the composition

* The properties of the quantities $-\psi$ and $-\zeta$ regarded as functions of the temperature and volume, and temperature and pressure, respectively, the composition of the body being regarded as invariable, have been discussed by M. Massieu in a memoir entitled "Sur les fonctions caractéristiques des divers fluides et sur la théorie des vapeurs" (*Mém. Savants Étrang.*, t. xxii). A brief sketch of his method in a form slightly different from that ultimately adopted is given in *Comptes Rendus*, t. lxix (1869), pp. 858 and 1057, and a report on his memoir by M. Bertrand in *Comptes Rendus*, t. lxxi, p. 257. M. Massieu appears to have been the first to solve the problem of representing all the properties of a body of invariable composition which are concerned in reversible processes by means of a single function.

and thermodynamic state of any such body without regard to its size or form. The word *phase* has been chosen for this purpose. Such bodies as differ in composition or state are called different phases of the matter considered, all bodies which differ only in size and form being regarded as different examples of the same pnase. Phases which can exist together, the dividing surfaces being plane, in an equilibrium which does not depend upon passive resistances to change, are called *coexistent.*

The number of independent variations of which a system of coexistent phases is capable is $n+2-r$, where r denotes the number of phases, and n the number of independently variable components in the whole system. For the system of phases is completely specified by the temperature, the pressure, and the n potentials, and between these $n+2$ quantities there are r independent relations (one for each phase), which characterize the system of phases.

When the number of phases exceeds the number of components by unity, the system is capable of a single variation of phase. The pressure and all the potentials may be regarded as functions of the temperature. The determination of these functions depends upon the elimination of the proper quantities from the fundamental equations in p, t, μ_1, μ_2, etc. for the several members of the system. But without a knowledge of these fundamental equations, the values of the differential coefficients such as $\frac{dp}{dt}$ may be expressed in terms of the entropies and volumes of the different bodies and the quantities of their several components. For this end we have only to eliminate the differentials of the potentials from the different equations of the form (12) relating to the different bodies. In the simplest case, when there is but one component, we obtain the well-known formula

$$\frac{dp}{dt}=\frac{\eta'-\eta''}{v'-v''}=\frac{Q}{t(v''-v')},$$

in which v', v'', η', η'' denote the volumes and entropies of a given quantity of the substance in the two phases, and Q the heat which it absorbs in passing from one phase to the other.

It is easily shown that if the temperature of two coexistent phases of two components is maintained constant, the pressure is in general a maximum or minimum when the composition of the phases is identical. In like manner, if the pressure of the phases is maintained constant, the temperature is in general a maximum or minimum when the composition of the phases is identical. The series of simultaneous values of t and p for which the composition of two coexistent phases is identical separates those simultaneous values of t and p for which no coexistent phases are possible from those for which there are two pairs of coexistent phases.

If the temperature of three coexistent phases of three components is maintained constant, the pressure is in general a maximum or minimum when the composition of one of the phases is such as can be produced by combining the other two. If the pressure is maintained constant, the temperature is in general a maximum or minimum when the same condition in regard to the composition of the phases is fulfilled.

Stability of fluids.—A criterion of the stability of a homogeneous fluid, or of a system of coexistent fluid phases, is afforded by the expression

$$\epsilon - t'\eta + p'v - \mu_1'm_1 - \mu_2'm_2 \ldots - \mu_n'm_n, \tag{14}$$

in which the values of the accented letters are to be determined by the phase or system of phases of which the stability is in question, and the values of the unaccented letters by any other phase of the same components, the possible formation of which is in question. We may call the former constants, and the latter variables. Now if the value of the expression, thus determined, is always positive for any possible values of the variables, the phase or system of phases will be stable with respect to the formation of any new phases of its components. But if the expression is capable of a negative value, the phase or system is at least *practically* unstable. By this is meant that, although, strictly speaking, an infinitely small disturbance or change may not be sufficient to destroy the equilibrium, yet a very small change in the initial state will be sufficient to do so. The presence of a small portion of matter in a phase for which the above expression has a negative value will in general be sufficient to produce this result. In the case of a system of phases, it is of course supposed that their contiguity is such that the formation of the new phase does not involve any transportation of matter through finite distances.

The preceding criterion affords a convenient point of departure in the discussion of the stability of homogeneous fluids. Of the other forms in which the criterion may be expressed, the following is perhaps the most useful:—

If the pressure of a fluid is greater than that of any other phase of its independent variable components which has the same temperature and potentials, the fluid is stable with respect to the formation of any other phase of these components; but if its pressure is not as great as that of some such phase, it will be practically unstable.

Stability of fluids with respect to continuous changes of phase.—In considering the changes which may take place in any mass, we have often to distinguish between infinitesimal changes in existing phases, and the formation of entirely new phases. A phase of a fluid may be stable with respect to the former kind of change, and unstable with respect to the latter. In this case, it may be capable of continued existence in virtue of properties which prevent the commencement of discontinuous changes. But a phase which is unstable with respect to continuous changes is evidently incapable of permanent existence on a large scale except in consequence of passive resistances to change. To obtain the conditions of stability with respect to continuous changes, we have only to limit the application of the variables in (14) to phases adjacent to the given phase. We obtain results of the following nature.

The stability of any phase with respect to continuous changes depends upon the same conditions with respect to the second and higher differential coefficients of the density of energy regarded as a function of the density of entropy and the densities of the several components, which would make the density of energy a minimum, if the necessary conditions with respect to the first differential coefficients were fulfilled.

Again, it is necessary and sufficient for the stability with respect to continuous changes of all the phases within any given limits, that within those limits the same conditions should be fulfilled with respect to the second and higher differential coefficients of the pressure regarded as a function of the temperature and the several potentials, which would make the pressure a minimum, if the necessary conditions with respect to the first differential coefficients were fulfilled.

The equation of the limits of stability with respect to continuous changes may be written

$$\left(\frac{d\mu_n}{d\gamma_n}\right)_{t,\,\mu_1,\,\ldots\,\mu_{n-1}}=0, \quad \text{or} \quad \left(\frac{d^2p}{d\mu_n^2}\right)_{t,\,\mu_1,\,\ldots\,\mu_{n-1}}=\infty, \tag{15}$$

where γ_n denotes the density of the component specified or $m_n \div v$. It is in general immaterial to what component the suffix $_n$ is regarded as relating.

Critical phases.—The variations of two coexistent phases are sometimes limited by the vanishing of the difference between them. Phases at which this occurs are called *critical phases.* A critical phase, like any other, is capable of $n+1$ independent variations, n denoting the number of independently variable components. But when subject to the condition of remaining a critical phase, it is capable of only $n-1$ independent variations. There are therefore two independent equations which characterize critical phases. These may be written

$$\left(\frac{d\mu_n}{d\gamma_n}\right)_{t,\,\mu_1,\,\ldots\,\mu_{n-1}}=0, \quad \left(\frac{d^2\mu_n}{d\gamma_n^2}\right)_{t,\,\mu_1,\,\ldots\,\mu_{n-1}}=0. \tag{16}$$

It will be observed that the first of these equations is identical with the equation of the limit of stability with respect to continuous changes. In fact, stable critical phases are situated at that limit. They are also situated at the limit of stability with respect to discontinuous changes. These limits are in general distinct, but touch each other at critical phases.

Geometrical illustrations.—In an earlier paper,* the author has described a method of representing the thermodynamic properties of substances of invariable composition by means of surfaces. The volume, entropy, and energy of a constant quantity of the substance are represented by rectangular coordinates. This method corresponds to the first kind of fundamental equation described above. Any other kind of fundamental equation for a substance of invariable composition will suggest an analogous geometrical method. In the present paper, the method in which the coordinates represent temperature, pressure, and the potential, is briefly considered. But when the composition of the body is variable, the fundamental equation cannot be completely represented by any surface or finite number of surfaces. In the case of three components, if we regard the temperature and pressure as constant, as well as the total quantity of matter, the relations between ζ, m_1, m_2, m_3 may be represented by a surface in which the distances of a point from the three sides of a triangular prism represent the quantities m_1, m_2, m_3, and the distance of the point from the base of the prism represents the quantity ζ. In the case of two components, analogous relations may be represented by a plane curve. Such methods are especially useful for illustrating the combinations and separations of the components, and the changes in states of aggregation, which take place when the substances are exposed in varying proportions to the temperature and pressure considered.

Fundamental equations of ideal gases and gas-mixtures.—From the physical properties which we attribute to ideal gases, it is easy to deduce their fundamental equations. The fundamental equation in ϵ, η, v, and m for an ideal gas is

$$c \log \frac{\epsilon - \mathrm{E}m}{cm} = \frac{\eta}{m} - \mathrm{H} + a \log \frac{m}{v}; \tag{17}$$

that in ψ, t, v, and m is

$$\psi = \mathrm{E}m + mt\left(c - \mathrm{H} - c \log t + a \log \frac{m}{v}\right); \tag{18}$$

that in p, t, and μ is

$$p = ae^{\frac{\mathrm{H}-c-a}{a}} t^{\frac{c+a}{a}} e^{\frac{\mu - \mathrm{E}}{at}}, \tag{19}$$

where e denotes the base of the Naperian system of logarithms. As for the other constants, c denotes the specific heat of the gas at

* [Page 33 of this volume.]

constant volume, a denotes the constant value of $pv \div mt$, E and H depend upon the zeros of energy and entropy. The two last equations may be abbreviated by the use of different constants. The properties of fundamental equations mentioned above may easily be verified in each case by differentiation.

The law of Dalton respecting a mixture of different gases affords a point of departure for the discussion of such mixtures and the establishment of their fundamental equations. It is found convenient to give the law the following form:—

The pressure in a mixture of different gases is equal to the sum of the pressures of the different gases as existing each by itself at the same temperature and with the same value of its potential.

A mixture of ideal gases which satisfies this law is called an *ideal gas-mixture.* Its fundamental equation in p, t, μ_1, μ_2, etc. is evidently of the form

$$p = \Sigma_1 \left(a_1 e^{\frac{H_1 - c_1 - a_1}{a_1}} t^{\frac{c_1 + a_1}{a_1}} e^{\frac{\mu_1 - E_1}{a_1 t}} \right), \tag{20}$$

where Σ_1 denotes summation with respect to the different components of the mixture. From this may be deduced other fundamental equations for ideal gas-mixtures. That in ψ, t, v, m_1, m_2, etc. is

$$\psi = \Sigma_1 \left(E_1 m_1 + m_1 t \left(c_1 - H_1 - c_1 \log t + a_1 \log \frac{m_1}{v} \right) \right). \tag{21}$$

Phases of dissipated energy of ideal gas-mixtures.—When the proximate components of a gas-mixture are so related that some of them can be formed out of others, although not necessarily in the gas-mixture itself at the temperatures considered, there are certain phases of the gas-mixture which deserve especial attention. These are the *phases of dissipated energy,* i.e., those phases in which the energy of the mass has the least value consistent with its entropy and volume. An atmosphere of such a phase could not furnish a source of mechanical power to any machine or chemical engine working within it, as other phases of the same matter might do. Nor can such phases be affected by any catalytic agent. A *perfect catalytic agent* would reduce any other phase of the gas-mixture to a phase of dissipated energy. The condition which will make the energy a minimum is that the potentials for the proximate components shall satisfy an equation similar to that which expresses the relation between the units of weight of these components. For example, if the components were hydrogen, oxygen and water, since one gram of hydrogen with eight grams of oxygen are chemically equivalent to nine grams of water, the potentials for these substances in a phase of dissipated energy must satisfy the relation

$$\mu_H + 8\mu_O = 9\mu_W.$$

Gas-mixtures with convertible components.—The theory of the phases of dissipated energy of an ideal gas-mixture derives an especial interest from its possible application to the case of those gas-mixtures in which the chemical composition and resolution of the components can take place in the gas-mixture itself, and actually does take place, so that the quantities of the proximate components are entirely determined by the quantities of a smaller number of ultimate components, with the temperature and pressure. These may be called *gas-mixtures with convertible components.* If the general laws of *ideal* gas-mixtures apply in any such case, it may easily be shown that the phases of dissipated energy are the only phases which can exist. We can form a fundamental equation which shall relate solely to these phases. For this end, we first form the equation in p, t, μ_1, μ_2, etc. for the gas-mixture, regarding its proximate components as *not* convertible. This equation will contain a potential for every proximate component of the gas-mixture. We then eliminate one (or more) of these potentials by means of the relations which exist between them in virtue of the convertibility of the components to which they relate, leaving the potentials which relate to those substances which naturally express the ultimate composition of the gas-mixture.

The validity of the results thus obtained depends upon the applicability of the laws of ideal gas-mixtures to cases in which chemical action takes place. Some of these laws are generally regarded as capable of such application, others are not so regarded. But it may be shown that in the very important case in which the components of a gas are convertible at certain temperatures, and not at others, the theory proposed may be established without other assumptions than such as are generally admitted.

It is, however, only by experiments upon gas-mixtures with convertible components, that the validity of any theory concerning them can be satisfactorily established.

The vapor of the peroxide of nitrogen appears to be a mixture of two different vapors, of one of which the molecular formula is double that of the other. If we suppose that the vapor conforms to the laws of an ideal gas-mixture in a state of dissipated energy, we may obtain an equation between the temperature, pressure, and density of the vapor, which exhibits a somewhat striking agreement with the results of experiment.

Equilibrium of stressed solids.—The second part of the paper* commences with a discussion of the conditions of internal and external equilibrium for solids in contact with fluids with regard to all possible states of strain of the solids. These conditions are deduced by

* [See footnote, p. 184.]

analytical processes from the general condition of equilibrium (2). The condition of equilibrium which relates to the dissolving of the solid at a surface where it meets a fluid may be expressed by the equation

$$\mu_1 = \frac{\epsilon - t\eta + pv}{m}, \tag{22}$$

where ϵ, η, v, and m_1 denote respectively the energy, entropy, volume, and mass of the solid, if it is homogeneous in nature and state of strain,—otherwise, of any small portion which may be treated as thus homogeneous,—μ_1 the potential in the fluid for the substance of which the solid consists, p the pressure in the fluid and therefore one of the principal pressures in the solid, and t the temperature. It will be observed that when the pressure in the solid is isotropic, the second member of this equation will represent the potential in the solid for the substance of which it consists {see (9)}, and the condition reduces to the equality of the potential in the two masses, just as if it were a case of two fluids. But if the stresses in the solid are not isotropic, the value of the second member of the equation is not entirely determined by the nature and state of the solid, but has in general three different values (for the same solid at the same temperature, and in the same state of strain) corresponding to the three principal pressures in the solid. If a solid in the form of a right parallelopiped is subject to different pressures on its three pairs of opposite sides by fluids in which it is soluble, it is in general necessary for equilibrium that the composition of the fluids shall be different.

The *fundamental equations* which have been described above are limited, in their application to solids, to the case in which the stresses in the solid are isotropic. An example of a more general form of fundamental equation for a solid, is afforded by an equation between the energy and entropy of a given quantity of the solid, and the quantities which express its state of strain, or by an equation between ψ {see (3)} as determined for a given quantity of the solid, the temperature, and the quantities which express the state of strain.

Capillarity.—The solution of the problems which precede may be regarded as a first approximation, in which the peculiar state of thermodynamic equilibrium about the surfaces of discontinuity is neglected. To take account of the condition of things at these surfaces, the following method is used. Let us suppose that two homogeneous fluid masses are separated by a surface of discontinuity, i.e., by a very thin non-homogeneous film. Now we may imagine a state of things in which each of the homogeneous masses extends without variation of the densities of its several components, or of the densities of energy and entropy, quite up to a geometrical surface (to be called the dividing surface) at which the masses meet. We may suppose this surface to be sensibly coincident with the physical surface of discontinuity. Now if we compare the actual state of things with the supposed state, there will be in the former in the vicinity of the surface a certain (positive or negative) excess of energy, of entropy, and of each of the component substances. These quantities are denoted by ϵ^{S}, η^{S}, m_1^{S}, m_2^{S}, etc., and are treated as belonging to the surface. The ^{S} is used simply as a distinguishing mark, and must not be taken for an algebraic exponent.

It is shown that the conditions of equilibrium already obtained relating to the temperature and the potentials of the homogeneous masses, are not affected by the surfaces of discontinuity, and that the complete value of $\delta\epsilon^{\text{S}}$ is given by the equation

$$\delta\epsilon^{\text{S}} = t\,\delta\eta^{\text{S}} + \sigma\,\delta s + \mu_1\delta m_1^{\text{S}} + \mu_2\delta m_2^{\text{S}} + \text{etc.}, \tag{23}$$

in which s denotes the area of the surface considered, t the temperature, μ_1, μ_2, etc., the potentials for the various components in the adjacent masses. It may be, however, that some of the components are found only at the surface of discontinuity, in which case the letter μ with the suffix relating to such a substance denotes, as the equation shows, the rate of increase of energy at the surface per unit of the substance added, when the entropy, the area of the surface, and the quantities of the other components are unchanged. The quantity σ we may regard as defined by the equation itself, or by the following, which is obtained by integration :—

$$\epsilon^{\text{S}} = t\eta^{\text{S}} + \sigma s + \mu_1 m_1^{\text{S}} + \mu_2 m_2^{\text{S}} + \text{etc.} \tag{24}$$

There are terms relating to variations of the curvatures of the surface which might be added, but it is shown that we can give the dividing surface such a position as to make these terms vanish, and it is found convenient to regard its position as thus determined. It is always sensibly coincident with the physical surface of discontinuity. (Yet in treating of plane surfaces, this supposition in regard to the position of the dividing surface is unnecessary, and it is sometimes convenient to suppose that its position is determined by other considerations.)

With the aid of (23), the remaining condition of equilibrium for contiguous homogeneous masses is found, viz.,

$$\sigma(c_1 + c_2) = p' - p'', \tag{25}$$

where p', p'' denote the pressures in the two masses, and c_1, c_2 the principal curvatures of the surface. Since this equation has the same form as if a tension equal to σ resided at the surface, the quantity σ is called (as is usual) the *superficial tension*, and the dividing surface in the particular position above mentioned is called the *surface of tension*.

By differentiation of (24) and comparison with (23), we obtain

$$d\sigma = -\eta_{\text{S}}dt - \Gamma_1 d\mu_1 - \Gamma_2 d\mu_2 - \text{etc.}, \tag{26}$$

where η_S, Γ_1, Γ_2, etc. are written for $\frac{\eta^S}{s}$, $\frac{m_1^S}{s}$, $\frac{m_2^S}{s}$, etc., and denote the superficial densities of entropy and of the various substances. We may regard σ as a function of t, μ_1, μ_2, etc., from which if known η_S, Γ_1, Γ_2, etc. may be determined in terms of the same variables. An equation between σ, t, μ_1, μ_2, etc. may therefore be called a *fundamental equation for the surface of discontinuity.* The same may be said of an equation between ϵ^S, η^S, s, m_1^S, m_2^S, etc.

It is necessary for the stability of a surface of discontinuity that its tension shall be as small as that of any other surface which can exist between the same homogeneous masses with the same temperature and potentials. Besides this condition, which relates to the nature of the surface of discontinuity, there are other conditions of stability, which relate to the possible motion of such surfaces. One of these is that the tension shall be positive. The others are of a less simple nature, depending upon the extent and form of the surface of discontinuity, and in general upon the whole system of which it is a part. The most simple case of a system with a surface of discontinuity is that of two coexistent phases separated by a spherical surface, the outer mass being of indefinite extent. When the interior mass and the surface of discontinuity are formed entirely of substances which are components of the surrounding mass, the equilibrium is always unstable; in other cases, the equilibrium may be stable. Thus, the equilibrium of a drop of water in an atmosphere of vapor is unstable, but may be made stable by the addition of a little salt. The analytical conditions which determine the stability or instability of the system are easily found, when the temperature and potentials of the system are regarded as known, as well as the fundamental equations for the interior mass and the surface of discontinuity.

The study of surfaces of discontinuity throws considerable light upon the subject of the stability of such phases of fluids as have a less pressure than other phases of the same components with the same temperature and potentials. Let the pressure of the phase of which the stability is in question be denoted by p', and that of the other phase of the same temperature and potentials by p''. A spherical mass of the second phase and of a radius determined by the equation

$$2\,\sigma = (p'' - p')r, \tag{27}$$

would be in equilibrium with a surrounding mass of the first phase. This equilibrium, as we have just seen, is unstable, when the surrounding mass is indefinitely extended. A spherical mass a little larger would tend to increase indefinitely. The work required to form such a spherical mass, by a reversible process, in the interior of an infinite mass of the other phase, is given by the equation

$$\mathrm{W} = \sigma s - (p'' - p')v''. \tag{28}$$

The term σs represents the work spent in forming the surface, and the term $(p''-p')v''$ the work gained in forming the interior mass. The second of these quantities is always equal to two-thirds of the first. The value of W is therefore positive, and the phase is in strictness stable, the quantity W affording a kind of measure of its stability. We may easily express the value of W in a form which does not involve any geometrical magnitudes, viz.,

$$W = \frac{16\pi\sigma^3}{3(p''-p')^2}, \tag{29}$$

where p'', p' and σ may be regarded as functions of the temperature and potentials. It will be seen that the stability, thus measured, is infinite for an infinitesimal difference of pressures, but decreases very rapidly as the difference of pressures increases. These conclusions are all, however, practically limited to the case in which the value of r, as determined by equation (27), is of sensible magnitude.

With respect to the somewhat similar problem of the stability of the surface of contact of two phases with respect to the formation of a new phase, the following results are obtained. Let the phases (supposed to have the same temperature and potentials) be denoted by A, B, and C; their pressures by p_A, p_B and p_C; and the tensions of the three possible surfaces by σ_{AB}, σ_{BC}, σ_{AC}. If p_C is less than

$$\frac{\sigma_{BC}\, p_A + \sigma_{AC}\, p_B}{\sigma_{BC} + \sigma_{AC}},$$

there will be no tendency toward the formation of the new phase at the surface between A and B. If the temperature or potentials are now varied until p_C is equal to the above expression, there are two cases to be distinguished. The tension σ_{AB} will be either equal to $\sigma_{AC}+\sigma_{BC}$ or less. (A greater value could only relate to an unstable and therefore unusual surface.) If $\sigma_{AB}=\sigma_{AC}+\sigma_{BC}$, a farther variation of the temperature or potentials, making p_C greater than the above expression, would cause the phase C to be formed at the surface between A and B. But if $\sigma_{AB} < \sigma_{AC}+\sigma_{BC}$, the surface between A and B would remain stable, but with rapidly diminishing stability, after p_C has passed the limit mentioned.

The conditions of stability for a line where several surfaces of discontinuity meet, with respect to the possible formation of a new surface, are capable of a very simple expression. If the surfaces A-B, B-C, C-D, D-A, separating the masses A, B, C, D, meet along a line, it is necessary for equilibrium that their tensions and directions at any point of the line should be such that a quadrilateral α, β, γ, δ may be formed with sides representing in direction and length the normals and tensions of the successive surfaces. For the stability of the system with reference to the possible formation of surfaces between A and C, or between B and D, it is farther necessary that the tensions σ_{AC} and σ_{BD} should be greater than the diagonals $\alpha\gamma$ and $\beta\delta$ respectively. The conditions of stability are entirely analogous in the case of a greater number of surfaces. For the conditions of stability relating to the formation of a new phase at a line in which three surfaces of discontinuity meet, or at a point where four different phases meet, the reader is referred to the original paper.

Liquid films.—When a fluid exists in the form of a very thin film between other fluids, the great inequality of its extension in different directions will give rise to certain peculiar properties, even when its thickness is sufficient for its interior to have the properties of matter in mass. The most important case is where the film is liquid and the contiguous fluids are gaseous. If we imagine the film to be divided into elements of the same order of magnitude as its thickness, each element extending through the film from side to side, it is evident that far less time will in general be required for the attainment of approximate equilibrium between the different parts of any such element and the contiguous gases than for the attainment of equilibrium between all the different elements of the film.

There will accordingly be a time, commencing shortly after the formation of the film, in which its separate elements may be regarded as satisfying the conditions of internal equilibrium, and of equilibrium with the contiguous gases, while they may not satisfy all the conditions of equilibrium with each other. It is when the changes due to this want of complete equilibrium take place so slowly that the film appears to be at rest, except so far as it accommodates itself to any change in the external conditions to which it is subjected, that the characteristic properties of the film are most striking and most sharply defined. It is from this point of view that these bodies are discussed. They are regarded as satisfying a certain well-defined class of conditions of equilibrium, but as not satisfying at all certain other conditions which would be necessary for complete equilibrium, in consequence of which they are subject to gradual changes, which ultimately determine their rupture.

The elasticity of a film (i.e., the increase of its tension when extended) is easily accounted for. It follows from the general relations given above that when a film has more than one component, those components which diminish the tension will be found in greater proportion on the surfaces. When the film is extended, there will not be enough of these substances to keep up the same volume- and surface-densities as before, and the deficiency will cause a certain increase of tension. It does not follow that a thinner film has always a greater tension than a thicker formed of the same liquid. When the phases within the films as well as without are the same, and the surfaces of the films are also the same, there will be no difference of tension. Nor will the tension of the same film be altered, if a part of the interior drains away in the course of time, without affecting the surfaces. If the thickness of the film is reduced by evaporation, its tension may be either increased or diminished, according to the relative volatility of its different components.

Let us now suppose that the thickness of the film is reduced until the limit is reached at which the interior ceases to have the properties of matter in mass. The elasticity of the film, which determines its stability with respect to extension and contraction, does not vanish at this limit. But a certain kind of instability will generally arise, in virtue of which inequalities in the thickness of the film will tend to increase through currents in the interior of the film. This probably leads to the destruction of the film, in the case of most liquids. In a film of soap-water, the kind of instability described seems to be manifested in the breaking out of the black spots. But the sudden diminution in thickness which takes place in parts of the film is arrested by some unknown cause, possibly by viscous or gelatinous properties, so that the rupture of the film does not necessarily follow.

Electromotive force.—The conditions of equilibrium may be modified by electromotive force. Of such cases a galvanic or electrolytic cell may be regarded as the type. With respect to the potentials for the ions and the electrical potential the following relation may be noticed:—

When all the conditions of equilibrium are fulfilled in a galvanic or electrolytic cell, the electromotive force is equal to the difference in the values of the potential for any ion at the surfaces of the electrodes multiplied by the electro-chemical equivalent of that ion, the greater potential of an anion being at the same electrode as the greater electrical potential, and the reverse being true of a cation.

The relation which exists between the electromotive force of a *perfect electro-chemical apparatus* (i.e., a galvanic or electrolytic cell which satisfies the condition of reversibility), and the changes in the cell which accompany the passage of electricity, may be expressed by the equation

$$d\epsilon = (V' - V'')de + t\,d\eta + dW_G + dW_P, \tag{30}$$

in which $d\epsilon$ denotes the increment of the intrinsic energy in the apparatus, $d\eta$ the increment of entropy, de the quantity of electricity which passes through it, V' and V'' the electrical potentials in pieces of the same kind of metal connected with the anode and cathode respectively, dW_G the work done by gravity, and dW_P the work done by the pressures which act on the external surface of the apparatus. The term dW_G may generally be neglected. The same is true of dW_P, when gases are not concerned. If no heat is supplied or withdrawn the term $t\,d\eta$ will vanish. But in the calculation of electromotive forces, which is the most important application of the equation, it is convenient and customary to suppose that the temperature is maintained constant. Now this term $t\,d\eta$, which represents the heat absorbed by the cell, is frequently neglected in the consideration of cells of which the temperature is supposed to remain constant. In other words, it is frequently assumed that neither heat or cold is produced by the passage of an electrical current through a perfect electro-chemical apparatus (except that heat which may be indefinitely diminished by increasing the time in which a given quantity of electricity passes), unless it be by processes of a secondary nature, which are not immediately or necessarily connected with the process of electrolysis.

That this assumption is incorrect is shown by the electromotive force of a gas battery charged with hydrogen and nitrogen, by the currents caused by differences in the concentration of the electrolyte, by electrodes of zinc and mercury in a solution of sulphate of zinc, by *a priori* considerations based on the phenomena exhibited in the direct combination of the elements of water or of hydrochloric acid, by the absorption of heat which M. Favre has in many cases observed in a galvanic or electrolytic cell, and by the fact that the solid or liquid state of an electrode (at its temperature of fusion) does not affect the electromotive force.

12

INVESTIGATION INTO THE FOUNDATIONS OF THERMODYNAMICS

C. Carathéodory

This article was translated expressly for this Benchmark volume by J. Kestin, Brown University, from "Untersuchungen über die Grundlagen der Thermodynamik," in Math. Ann. (Berlin), **67**, *355–386 (1909)*

Introduction

The proposition that the discipline of thermodynamics can be justified without recourse to any hypothesis that cannot be verified experimentally must be regarded as one of the most noteworthy results of the research into thermodynamics that was accomplished during the last century. The point of view adopted by most authors who were active in the last fifty years—that is, after the great discoveries of R. Mayer—the measurements of Joule and the fundamental work of Clausius and W. Thomson is, essentially, the following:

There exists a physical quantity called heat that is not identical with the mechanical quantities (mass, force, pressure, etc.) and whose variations can be determined by calorimetric measurements. In certain circumstances, heat has the property that it can become equivalent to the ordinary mechanical work. Furthermore, when two bodies of different temperatures are brought into contact, heat always passes from the hotter to the colder, and never in the reverse direction.

Even though no further assumptions regarding the nature of heat were made, it was possible to construct a theory the results of which always agree with experience. The understanding of this theory was later facilitated by the introduction of a new quantity, the energy, whose importance for the whole of physics revealed itself slowly with the passage of time. This physical quantity has the property that it depends only on the instantaneous state of the various substances under consideration. This is an advantage that is not shared by heat.

The first law of thermodynamics has now transformed itself into a *definition of energy* and states that this quantity can always be determined with the aid of mechanical and calorimetric measurements.

Now, several authors have already noticed that such a view hides an element of overdetermination.[1] *It is possible to develop the whole theory without assuming the existence of heat, that is, of a quantity that is of a different nature than the normal mechanical quantities.*

[1] Bryan, "Thermodynamics," *Enzyklopädie der mathem. Wiss. V. 3*, p. 81; J. Perrin, "Le contenu essentiel des principes de la Thermodynamique," *Bull. de la soc. franç. de philos.* V.VI, 1906, p. 81.

To present all this in all detail and as clearly as possible is the purpose of the present paper. It is, naturally, possible to construct a physical theory in very different ways. I have chosen an arrangement of conclusions that differs as little as possible from classical proofs and yet allows us to recognize the parallelism that must exist between the assertions of the theory and the pictures presented by actual measurements.

The essential characteristic of the presentation given here consists in the fact that the concepts "adiabatic" and "abiabatically insulated" are defined with the aid of physical properties and are not, as is usual, reduced to the concept of energy. In such circumstances, it becomes possible to state the axiom of the first law so that it corresponds exactly to Joule's experimental arrangement if we regard the calorimeter employed in the experiment as an adiabatically isolated system.

As far as the axiom of the second law is concerned, I have chosen a definition that is closely related to Planck's; however, the latter had to be suitably modified in order to account for the fact that the concepts of heat and quantity of heat have not yet been defined.

I have examined in detail the conditions under which an adiabatic process becomes reversible or, more precisely, a set of sufficient conditions for this to be so. In this manner I arrived at the definition of certain thermodynamic systems that can be called "simple," because they can be treated in exactly the same way as the simplest systems known in thermodynamics. These conditions deviate from those that have been introduced by Bryan in his Encyclopedia article[2] cited earlier.

Further, it became necessary to make use of a theorem of the theory of Pfaffian differential equations for which I give a simple proof in the fourth section. This was done in order to enable us to consider systems endowed with an arbitrary number of degrees of freedom from the outset, rather than to refer to the properties of Carnot cycles that are still widely used, and that can be clearly and easily mastered only in relation to systems with two degrees of freedom.

Finally, I wish to draw attention to the fact that the concept of temperature has not been included among the coordinates from the very beginning. Instead, I have defined it as a result of certain conditions that will be established on page 28 (**244**). The reasons for preferring this formulation for temperature are briefly sketeched in the concluding section; they result from certain considerations that have been originated from an analysis of radiation.

1. DEFINITIONS

The succeeding investigation is concerned with the description of the thermal properties of systems that consist of different chemical substances.

The general principles that will serve us to attain such a description appear in their complete generality when we consider, for the sake of brevity, a more restricted problem and make the same assumptions as those adopted by Gibbs in the first part of his fundamental treatise "On the Equilibrium of Heterogenous Substances."[3]

[2] Loc. cit, p. 80.

[3] J. W. Gibbs, *Scientific Papers*, Vol. 1, p. 55. [*Editor's Note*: See also pp. **211–228** of this book.]

We shall indicate at the end of this paper how further questions can be analyzed with the aid of the same principles.

Thus, following Gibbs,[4] we shall postulate that there exist systems S that, when they are present in equilibrium, consist of a finite number α of liquid or gaseous homogeneous parts—the "phases"

$$\phi_1, \phi_2, \ldots, \phi_\alpha$$

of the system—and that forces acting at a distance, such as gravity, as well as electromagnetic and capillary forces, can be neglected.[5]

Thus, the systems S that we wish to consider are *defined* when certain properties are associated with them in such a way that they characterize them completely.

For this purpose we consider an arbitrary equilibrium state of S and examine its phases

$$\phi_1, \phi_2, \ldots, \phi_\alpha$$

one by one. With each phase ϕ_i we associate two kinds of properties: first certain *characteristics* that define the chemical composition of ϕ_i so that we can enumerate the substances and compounds that occur in ϕ_i; secondly, we indicate *numerical values* that are obtained with the aid of *measurements.* These numbers represent the following quantities:

a. the total volume v_i of phase ϕ_i

b. the pressure p_i exerted by the phase under consideration on the contiguous bodies

c. the amounts

$$m_{1i}, m_{2i}, \ldots, m_{\beta i}$$

of the various substances and compounds that exist in every unit volume of ϕ_i.

For example, if the first phase ϕ_1 consists of a solution of common salt in water, we shall suppose that a given equilibrium state of this phase is completely defined for our theory with the aid of the numbers V_1, p_1, and a symbolic equation, such as

$$\phi_1 = m_{11}\ H_2O + m_{21}\ NaCl. \tag{1}$$

All phases are completely characterized by symbolic equations like Eq. (1) and by the set of numbers

$$v_i, p_i, m_{\kappa i}; \quad i = 1, 2, \ldots, \alpha; \quad \kappa = 1, 2, \ldots, \beta. \tag{2}$$

However, the whole system S has not yet been characterized completely, because

[4] Loc. cit., p. 62.

[5] The consequences of the last two assumptions do not come into play until the end of the second section.

we must take into account the properties of S that arise as a result of the contact between the phases among themselves on the one hand, and between the phases and the walls of the vessels, on the other.

We assume that the mass of the walls is so small that our results will remain unaffected by the fact that we have not included the walls themselves among the phases ϕ_1. Such a restriction would automatically disappear in a more general theory in which solid isotropic and solid crystalline bodies can also be counted among the phases.

The physical properties of the walls of the vessels that contain one or more phases occur in many forms.

A vessel Γ of this kind can have, for example, the property that the phases present in it are in equilibrium and that the numerical values (2) established for them remain constant when the bodies present outside the vessel are modified with the single restriction that Γ remains at rest and retains its original shape. A thermos flask constitutes a handy example of such a vessel. However, I wish explicitly to stress that the walls of Γ need not be rigid, because it is possible to imagine a perfectly deformable vessel Γ that possesses the preceding property; in such cases, the changes in the bodies present outside must be restricted so that the pressures exerted on Γ do not deform it.

A vessel provided with such properties is called *adiabatic* and the phases enclosed in it are referred to as *adiabatically isolated.*

If two phases, ϕ_1 and ϕ_2, remain in contact across a rigid adiabatic wall, it is possible to assert, in analogy with the above definition, that such a contact does not impose a constraining equation between V_1, p_1, m_{K1} and V_2, p_2, m_{K2}; in other words, two arbitrary equilibrium states of two phases of S can coexist on two sides of the wall.

In the case of other rigid walls, equilibrium can exist only on condition that one or more relations of the form

$$F(V_1, p_1, m_{K1};\quad V_2, p_2, m_{K2}) = 0 \tag{3}$$

are satisfied. It is then said that the wall is "*permeable.*" A wall can be permeable only "to heat" or, in addition, to some chemical substances with which it is in contact; more complex relationships can also arise. In every particular case it is necessary to *define* what is meant by the various expressions used. This is done *experimentally* by establishing the form of an equation of constraint of type (3) that describes the thermodynamic properties of the wall under consideration. When the walls are also deformable, it is necessary to add the condition that the pressure is the same on both sides.

When two phases, ϕ_1 and ϕ_2, remain in contact in the absence of a wall, the necessary conditions (3) for equilibrium continue to apply.

The discovery of all relations of this type constitutes one of the principal problems of experimental thermodynamics; we shall examine the most important ones among them further along in this paper.

Let us now denote the numerical values (2) uniformly by

$$c_0, c_1, \ldots, c_{n+\lambda}. \tag{4}$$

For the moment, it is sufficient to know that they must satisfy a number of mutually independent equations

$$
\begin{aligned}
F_1(c_0, c_1, \ldots, c_{n+\lambda}) &= 0 \\
F_2(c_0, c_1, \ldots, c_{n+\lambda}) &= 0 \\
&\cdot \quad \cdot \quad \cdot \\
F_\lambda(c_0, c_1, \ldots, c_{n+\lambda}) &= 0
\end{aligned} \tag{5}
$$

in order to secure the existence of equilibrium.

We now assume that it is possible to determine *all* equations of this kind that are valid for S. This means that we can establish experimentally equilibrium states for every combination of the numbers (4) that satisfy Equations (5). Experience teaches that this is possible in every concrete case.

Definition I. *Two systems S and S' are called "equivalent" when there exists between them a one-to-one relation between their phases in the sense of equation (1) and when, in addition, the corresponding coefficients c_i, c_i' must be subject to the same or to equivalent conditions (5) in order to secure equilibrium.*

In what follows, we shall make no distinction between equivalent systems. The properties that *define* our system S consist, therefore, of symbolic equations, like Eq. (1) on the one hand, and of a system of equations (5) on the other.

We now augment the system of equations (5) by $(n + 1)$ equations of the form

$$
\begin{aligned}
G_0(c_0, c_1, \ldots, c_{n+\lambda}) &= x_0 \\
G_1(c_0, c_1, \ldots, c_{n+\lambda}) &= x_1 \\
&\cdot \quad \cdot \quad \cdot \\
G_r(c_0, c_1, \ldots, c_{n+\lambda}) &= x_n.
\end{aligned} \tag{6}
$$

The functions G_i have been so chosen that there exists a one-to-one relation between the possible sets of numbers

$$c_0, c_1, \ldots, c_{n+\lambda}$$

and the corresponding values

$$x_0, x_1, \ldots, x_n$$

when the c_i are varied over a range of values—which actually occurs in practice—in such a way that conditions (5) remain satisfied. For this to be the case it is necessary (but not sufficient) that the Jacobian

$$\frac{\partial(G_0, G_1, \ldots, G_n; \quad F_1, F_2, \ldots, F_\lambda)}{\partial(c_0, c_1, \ldots, c_{n+\lambda})}$$

shall not vanish for all possible values of c_i.[6] If this is so, it becomes possible to solve the system of Eqs. (5), (6) for the c_i and to regard the c_i as the functions

$$c_i = c_i(x_0, x_1, \ldots, x_n), \tag{7}$$

which turn the system of equations into identities when substituted into (5).

The existence of the one-to-one relation between the various possible equilibrium states of S and the sets of values

$$x_0, x_1, \ldots, x_n \tag{8}$$

give us an analytic tool that can be used to compare these equilibrium states with each other and to represent them with the aid of "generalized coordinates" in analogy to those that are used in mechanics.

To every set of values (8) there corresponds what is called a "state" of the system S. As far as the numbers x_i themselves are concerned, we wish to introduce the term "state variables" for them.

In order to be able to use the language of geometry, it is now convenient to regard the state variables as Cartesian coordinates of a space of $(n + 1)$ dimensions. To every state of S there now corresponds a point in the multidimensional space, and the set of possible equilibrium states is uniquely mapped on a certain domain G of this space.

Summarizing, we can now state:

Definition II. *In order to characterize the equilibrium states of a system it is necessary to consider only the state variables (8). Two equivalent systems for which these quantities assume identical values represent identical objects from the point of view of thermodynamics.*

We now consider "processes" performed by the system S, that is, transitions from one equilibrium state to another. Processes, like equilibrium states, are characterized by certain properties. We shall take the latter as the state variables of the initial and final equilibrium state. In addition we must include a further quantity that is associated with every process and that is called external work.[7] We assume that this quantity, denoted by A, is produced entirely by the deformation in the *external* shape of S.[8] This work is identical with the mechanical work performed by the forces that are exerted by S on the external bodies (those in contact with S). The physical significance of these forces is clear. The work A can also be measured with the aid of suitable mechanical devices, as is the case in engineering when steam or gas engines are tested on a stand.

To conclude, we associate one more property with every process. If during a process the system S remains adiabatically isolated at all times, we shall call the process itself *adiabatic*. Adiabatic processes form a very special class.

[6] The fact that the functional determinant does not vanish guarantees merely the existence of a one-to-one relation between the c and the κ "in the small," that is, in the neighborhood of every single point; it does not assure us that the domains of the x do not overlap.

[7] [*Editor's Note*: This is actually called "internal work" in mechanics.]

[8] This is a consequence of the fact that we ignore forces acting at a distance. [*Editor's Note*: We also ignore the existence of internal deformation variables.]

In this manner we are led to the following definition:
Definition III. *Every process is characterized by the state variables of the initial and final state, by the external work performed by it, and by the indication whether it is adiabatic or not.*

2. AXIOMS

The concepts defined in the preceding section allow us to state certain *axioms*, that is, generalizations from experience that have been obtained in particularly simple circumstances. The science of thermodynamics knows two such mutually independent axioms.

The first axiom serves as a basis for the so-called first law of thermodynamics and is no more than an expression for the general principle of energy conservation applied to the systems under consideration. We wish to formulate it as follows:
Axiom I: *With every phase ϕ_i of a system S it is possible to associate a function ϵ_i of the quantities (2), i.e.,*

$$V_i, p_i, m_{\kappa i},$$

which is proportional to the total volume V_i of the phase and which is called its internal energy.

The sum

$$\epsilon = \epsilon_1 + \epsilon_2 + \ldots + \epsilon_\alpha$$

extended over all phases is called the internal energy of the whole system.

During an adiabatic process, the sum of the work A and the energy difference vanishes. Denoting the initial and final energy by ϵ and $\bar{\epsilon}$, respectively, we can write this as

$$\bar{\epsilon} - \epsilon + A = 0. \tag{9}$$

This formulation of the first law is consistent with the assumption made at the beginning of this paper, namely, that neither forces acting at a distance nor those of capillary action are to be included. For example, if capillary forces were to be included, the sum of the various volumetric energies of the phases would not be equal to the energy ϵ of S and it would be necessary to add to it various terms associated with the surfaces that separate the phases. Similary, new terms would arise if forces acting at a distance between the phases were to come into play. Since such forces arise as a result of interactions between phases, they would not each depend on a single phase but on several among them.

The second law that should now be introduced is entirely different in nature: It has, namely, been discovered that by performing all adiabatic processes that start with an arbitrarily given initial state it is not possible to reach certain end states and that such "inaccessible" final states can be found in every close neighborhood of the prescribed initial state.

Since, however, physical measurements can never be made with an absolute accuracy, it is necessary to realize that the preceding experimental fact expresses more than its bare mathematical content would suggest. For this reason, when a certain point is excluded, we must demand that the same must be true about a small region surrounding the point, the extent of that region being a function of experimental accuracy. In order not to be forced to take into account considerations regarding accuracy, it is useful to clothe the preceding axiom in somewhat more general terms, and to state:

Axiom II: *In every arbitrarily close neighborhood of a given initial state there exist states that cannot be approached arbitrarily closely by adiabatic processes.*

3. SIMPLE SYSTEMS

The object of our next step is to investigate, with the aid of the two laws, how it is possible experimentally to determine the internal energy of every system of interest and, simultaneoulsy, to discover the general properties of the energy function ϵ.

We shall see that these problems can be solved relatively easily for certain special systems, which we shall call "simple systems." Now, if we succeed in combining a given phase ϕ_1—whose energy we wish to establish—with several phases whose energies have been measured earlier, into such a simple system, we shall possess all the necessary data required; indeed, according to Axiom I we can write the equation

$$\epsilon_1 = \epsilon - \epsilon_2 - \epsilon_3 - \ldots - \epsilon_\alpha.$$

The problem of constructing a "simple system" for every given phase in every practical case that may be encountered constitutes one of the most important—and most difficult—problems of experimental thermodynamics. Physical chemists describe this process by the phrase "to render a process reversible." However, as far as our general considerations are concerned, this problem does not arise; it is enough to know that it can be mastered, generally speaking.

The properties that characterize "simple systems" are varied.

First, all state parameters of S, *except for one*, must depend on the external shape of the system. The coordinates that determine the external shape will be called *deformation variables*; taking into account Eq. (5), we see that they can contain only the state parameters $V_1, V_2, \ldots, V_\alpha$.

For example, this is the case when the system consists of a single phase in which all state parameters, except for the pressure p and the total volume V, remain constant. This is also true when the system consists of two such phases that are separated by a rigid wall that is permeable "only to heat." In this case, the external shape of S depends on two state parameters, namely, on the total volumes V_1 and V_2 of the two subsystems. The whole system is characterized by two deformation variables, and the four state variables under consideration—namely, V_1, p_1, V_2, p_2—must satisfy the relation

$$F(V_1, p_1, V_2, p_2) = 0$$

because the partition is permeable (see Section 6). In the last analysis, the system is characterized by three state variables.

This first property of simple systems makes it possible to determine the end state of the simple system S when the initial state in an adiabatic process is known together with the final shape and the work performed. Indeed, we may write

$$\bar{\epsilon} - \epsilon + A = 0, \tag{9}$$

and Eq. (9) allows us to complete the calculation assuming that the energy ϵ is known as a function of the state variables and that, as we shall see presently, it is not completely determined by the shape of the system.

A second assumption postulated for simple systems consists in the statement that a knowledge of the initial and final shape alone does not uniquely determine the external work A performed during an adiabatic process. On the contrary, we assume that it is possible to perform adiabatic processes that lead from a given initial state to the same prescribed final shape and that are characterized by different quantities of work performed. For example, if a gas is contained in an adiabatic cylinder fitted with a piston, the work performed by the latter will depend on the speed with which the piston is withdrawn, even if the volumetric expansion is given the same prescribed value in all cases.

This assumption leads to the conclusion that the energy ϵ, regarded as a function of the state variables x_i, must contain the variable that is independent of the shape of S. This can be inferred by reference to Eq. (9). We denote this state variable by x_0, and reserve the symbols $x_1, \ldots, x_n$ for the deformation variables of the system.

We now wish to consider all values of A that are possible when the system performs a process from a prescribed initial state to a prescribed final shape. The totality of these points can be regarded as a point-set on a line. We now assume a third property of simple systems in that we postulate that this set of points is *connected* in all cases. In other words, the points fill a single interval that, otherwise, could extend to infinity in both directions.

It follows from Eq. (9), and with the aid of this last property, that all possible values of x_0 also form a connected set of points under identical circumstances, at least in cases when the range of values of the state variables is restricted to a certain neighborhood of the initial state.

Starting with a given initial state, it is always possible to reach any final shape by the application of suitable external forces. Even more, during an adiabatic process it is possible to prescribe the change in the shape of system S as a function of time. In other words, it is possible to prescribe the functions

$$x_1(t), x_2(t), \ldots, x_n(t), \tag{10}$$

and to demand that the process occur in such a way that the temporal variations of the coordinates shall be described by the set (10). This new description of a process that leaves out of account only the time-variation of x_0 is much more detailed than the one discussed previously; however, we wish to leave open the question as to whether the initial state and the functions (10) uniquely determine the magnitude of the corresponding work A during an adiabatic process. On the other hand, when the rate with which the system is deformed becomes "infinitely slow," or, more precisely, when the derivatives

$$x_1'(t), x_2'(t), \ldots, x_n'(t)$$

converge to zero uniformly, the quantity of work A tends to a definite value in the limit. A process that occurs so slowly that the difference between the work performed externally and the preceding limit is smaller than the uncertainty of our measurements will be called "*quasistatic.*"

If, for a quasistatic adiabatic process, the external work is also known as a function of time, it becomes possible to regard the last state variable x_0 as a function of time t. This can be seen from the equation

$$\epsilon\left\{ x_0, x_1(t), \ldots, x_n(t) \right\} - \epsilon_0 + A(t) = 0,$$

in which ϵ_0 denotes the initial value of energy. Accordingly, a quasistatic adiabatic process can be regarded as a *series of equilibrium states,* and to every quasistatic reversible process there corresponds a specified curve in the space of the x_i.

To conclude this line of thought we adopt the last assumption: During any quasistatic reversible process, the external work A can be so determined as if the forces that produce it were equal to those required to maintain equilibrium. This is true whenever the process can be regarded as a series of equilibrium states. However, the latter forces are sole functions of the state variables.

According to the preceding, the expression for A must be of the form

$$A(t) = \int_{t_0}^{t} DA, \tag{11}$$

where DA denotes the Pfaffian form

$$DA = p_1\, dx_1 + p_2 + \ldots + p_n\, dx_n, \tag{12}$$

and the $p_1, \ldots, p_n$ are functions of the variables $x_0, x_1, \ldots, x_n$.

The functions p_i can be determined experimentally by measuring at every state of S those forces that must be applied from the outside to maintain equilibrium. As far as adiabatic quasistatic processes are concerned, Eq. (9) of the first law can now be written

$$\int_{t_0}^{t} (d\epsilon + DA) = 0, \tag{13}$$

and since this relation is valid for any t, it follows that only such curves of the $(n + 1)$-dimensional space of the x_i can represent quasistatic adiabatic processes as satisfy the Pfaffian equation

$$d\epsilon + DA = 0, \tag{14}$$

Inversely, every curve in the $(n + 1)$-dimensional space of the x_i that satisfies Eq. (14) constitutes the image of a quasistatic adiabatic process of our simple system.

Let us assume that such an element of curve is given in parametric form by the equations

$$x_0 = x_0(\tau), x_1 = x_1(\tau), \ldots, x_n(\tau)$$
$$0 \leqslant \tau \leqslant 1. \tag{15}$$

We put

$$\tau = \lambda\tau,$$

where t denotes the time and λ is a parameter. We can now introduce adiabatic processes that satisfy the equation

$$x_1 = x_1(\lambda t), x_2 = x_2(\lambda t), \ldots, x_n = x_n(\lambda t)$$

for any prescribed value of λ. For λ sufficiently small the process becomes quasistatic and must, according to what has been said earlier, satisfy Eq. (14). By integration, we see that

$$x_0 = x_0(\lambda t),$$

and this proves our assertion.

If we substituted

$$\tau = 1 - \lambda t$$

in Eqs. (15), we would discover that for sufficiently small values of λ exactly the same curve would be traversed, but in the reverse direction. In this manner we have proved the following proposition:

Quasistatic adiabatic processes of a simple system are "reversible."

In other presentations of the theory, the concept of a "reversible" process is introduced as something self-evident; however, when we examine the matter more closely, it appears that the properties that we assign to reversible processes are exactly those that we have used as a basis for the definition of our simple system. We summarize them as follows:

Definition: *A "simple" system endowed with $(n + 1)$ state variables $x_0, x_1, \ldots, x_n$ must satisfy the following conditions:*

1. n of its coordinates—for example, $x_1, x_2, \ldots, x_n$—are deformation coordinates.

2. The external work A is not uniquely determined by the initial and final state of S in an adiabatic process; the totality of all possible values in the given circumstances forms a connected set of numbers.

3. During "quasistatic" reversible processes the external work is equal to the integral of a definite Pfaffian expression of the form

$$DA = p_1\, dx_1 + p_2\, dx_2 + \ldots + p_n\, dx_n.$$

Normally it is accepted that the first assumption concerning the number of deformation coordinates implies also the other two. Thus, the examples given on p. 236 constitute simple systems. This assumption—and we are going to accept it from now on—is admissible for substances of normal interest; in particular, it is admissible for gases and liquids. This is due to the fact that the conclusions that we draw from the theory agree very well with results of measurements.

By contrast, it is possible to think, and it appears to be physically reasonable to assume, that there exist substances in nature that can never be regarded as components of simple systems. For example, this would be the case if the internal friction of the substance under consideration—which is, generally speaking, a function of the deformation rate—were not to converge to zero during quasistatic processes. In such cases, the forces that produce the external work A would cease to be comparable to the equilibrium forces: The work A itself could no longer be represented by a Pfaffian form as in (12), and, finally, quasistatic processes would no longer be reversible. Our theory cannot be extended to include such systems without further thought. The same can be said quite generally also about classical thermodynamics.

The application of the axiom of the second law of thermodynamics to quasistatic adiabatic processes of simple systems will now allow us to normalize the state variables of such systems in a particular way; however, in order to do this we need to apply a mathematical theorem on Pfaffian equations, which we now proceed to derive.

4. AUXILIARY THEOREM OF THE THEORY OF PFAFF'S EQUATIONS

Give a Pfaffian equation

$$dx_0 + X_1\,dx_1 + X_2\,dx_2 + \ldots + X_n\,dx_n = 0 \tag{16}$$

in which the X_i denote continuous differentiable functions of the x_i, and granted that in every neighborhood of a given point P of the space of x_i points there exist points that cannot be reached along curves that satisfy this equation, it necessarily follows that the expression (16) possesses a multiplier that turns it into a perfect differential.

Let

$$a_0, a_1, \ldots, a_n$$

be the coordinates of P. According to our assumption, there exist infinitely many points P_i for which point P is a point of accumulation and which have the property that there does not exist a single curve that satisfies Eq. (16) and that possess through P and P_i simultaneously.

Nevertheless, since the coefficient of x_0 does not vanish, it is always possible to find curves C_i that satisfy the differential equation (16), contain P_i, and lie in the two-dimensional plane that contains P_i and the straight line G given by

$$x_0 = t, \quad x_k = a_k \quad (k = 1, 2, \ldots, n),$$

if P_i itself does not already lie on this straight line. Let Q_i be the point of intersection of C_i with G. According to the way in which they have been constructed, the points Q_i must converge towards P as i increases. The points Q_i cannot be reached from P along curves that satisfy the differential equation (16). This is due to the fact that in the contrary case we could include curve C_i and reach P_i in violation of our assumption. It follows that in every interval on the straight line G that contains P there must exist points that are inaccessible from P.

Now let us consider a straight line G_1 that is parallel to G, but otherwise arbitrary. Let us construct a two-dimensional cylinder that connects G and G_1, and let M be the point at which the line G_1 is intersected by a curve with the following properties: The curve satisfies Eq. (16), lies on the cylinder, and contains P. As we deform the cylinder in an arbitrary fashion, point M must remain fixed; in the contrary case the curve (which satisfies the differential equation (16), lies on the deforming cylinder, and passes through M) could contain some point on G that lies in the neighborhood of P. In this manner it would become possible to reach certain points Q_i starting from P and proceeding via M along curves of Eq. (16).

Let us now continuously displace the straight line G_1; it is seen that M will trace an n-dimensional surface that necessarily contains all curves satisfying Eq. (16) and passing through P. However, point P was chosen arbitrarily; by varying its position we obtain a family of surfaces

$$P(x_0, x_1, \ldots, x_n) = C$$

that depend on the parameter C and that contain *all* curves of Eq. (16). The coefficients of the dx_i in the two equations, namely,

$$dx_0 + X_1\,dx_1 + X_2\,dx_2 + \ldots + X_n\,dx_n = 0$$

$$\frac{\partial F}{\partial x_0}\,dx_0 + \frac{\partial F}{\partial x_1}\,dx_1 + \frac{\partial F}{\partial x_2}\,dx_2 + \ldots + \frac{\partial F}{\partial x_n}\,dx_n = 0$$

must be proportional to each other, and we can write the relation

$$dF = \frac{\partial F}{\partial x_0}\left\{dx_0 + X_1\,dx_1 + X_2\,dx_2 + \ldots + X_n\,dx_n\right\}$$
$$\frac{\partial F}{\partial x_0} \neq 0; \quad \frac{\partial F}{\partial x_0} \neq \infty. \tag{17}$$

This proves our theorem.

5. NORMALIZATION OF THE COORDINATES OF A SIMPLE SYSTEM

Let S be a simple system whose state depends on the coordinates

$$\xi_0, x_1, x_2, \ldots, x_n, \tag{18}$$

where the last n quantities in the series (18) are deformation coordinates. Hence, the external work of a quasistatic adiabatic process is given by the integral of

$$DA = p_1\,dx_1 + p_2\,dx_2 + \ldots + p_n\,dx_n.$$

The adiabatic quasistatic processes of the system can be represented by the curves of the following Pfaffian equation:

$$\begin{aligned} d\epsilon + DA &= \frac{\partial\epsilon}{\partial\xi_0}\,d\xi_0 + X_1\,dx_1 + X_2\,dx_2 + \ldots + X_n\,dx_n = 0 \\ X_i &= \frac{\partial\epsilon}{\partial x_i} + p_i \end{aligned} \qquad (19)$$

Should it now be possible to reach every point of a certain neighborhood of an initial point moving along curves of the differential equation, then it would follow from our assumptions concerning simple systems that every arbitrary final state of such a system could be produced from a given initial state by an adiabatic quasi-static process. However, according to our Axiom II, this is not possible. On the other hand, it follows from the properties of simple systems that $\partial\epsilon/\partial\xi_0$ does not vanish identically. It thus becomes possible to divide the expression in (19) by $\partial\epsilon/\partial\xi_0$ (disregarding certain singular points) and so to reproduce exactly the assumptions of the preceding section.

It follows that the expression in (19) possesses a multiplier that is neither infinite nor zero; if we denote it by $1/M$, we obtain

$$d\epsilon + DA = M\,dx_0, \qquad (20)$$

where x_0 is a specified function of the variables (18). Upon comparing (19) with (20), we see that

$$\frac{\partial x_0}{\partial\xi_0} = \frac{\partial\epsilon/\partial\xi_0}{M}$$

is different from zero. This allows us to solve the equation

$$x_0 = x_0 \quad (\xi_0, x_1, x_2, \ldots, x_n)$$

for ξ_0 and to adopt x_0 as our $(n+1)$st coordinate instead of ξ_0, and to add it to the n deformation coordinates.

If we do this, the expression for work

$$DA = p_1\,dx_1 + p_2\,dx_2 + \ldots + p_n\,dx_n \qquad (21)$$

retains its original form, because it does not contain the differential $d\xi_0$. However, the p_i must now be expressed in terms of the new variables $x_0, x_1, \ldots, x_n$. Similarly, the functions M in (20) must be assumed expressed in terms of the same variables.

Curves that correspond to adiabatic quasistatic processes of our system satisfy the equation

$$x_0 = \text{Const.} \tag{22}$$

Reciprocally, every curve in the space of the x_i along which x_0 remains constant can be regarded as the image of such a process. This follows from the fact that Eq. (22) is equivalent to (19), and we have seen in the third section that curves that satisfy the latter equation do possess the required property.

Let us now observe that Eq. (20) constitutes an identity and let us substitute in it DA from (21). We thus obtain the relations

$$M\,dx_0 = d\epsilon + DA = \left(\frac{\partial \epsilon}{\partial x_0}\right) dx_0 \tag{23}$$

$$p_i = -\frac{\partial \epsilon}{\partial x_i} \qquad (i = 1, 2, \ldots, n). \tag{24}$$

A system of coordinates that possesses all of the properties introduced in (21), (23), (24) will now be called a "normalized" system of coordinates. In this connection it is necessary to notice that all these properties remain valid when x_0 is replaced by an arbitrary function $f(x_0)$ of the quantity. The same result follows directly from the theory of integrating factors of a Pfaffian expression.

In thermodynamics it is possible to identify one special normalized system of coordinates that can be made unique with the aid of the physical properties of rigid walls that are permeable only to heat; to do this is our next task.

6. CONDITIONS FOR THERMAL EQUILIBRIUM

Let there be given two simple systems, S_1 and S_2, with normalized coordinates

$$x_0, x_1, \ldots, x_n$$

$$y_0, y_1, \ldots, y_n.$$

Let the systems be separated by a rigid wall permeable to heat. Such a wall is defined with the aid of the following properties:

1. The deformation coordinates of both systems under consideration can be varied independently after the introduction of the coupling.
2. After every arbitrary manipulation of the shape of the total system, a state of equilibrium sets in after a finite time if the system is adiabatically isolated.
3. The total system S exists in a state of equilibrium when and only when the coordinates x_i, y_k satisfy a certain relation of the form

$$F(x_0, x_1, \ldots, x_n;\ y_0, y_1, \ldots, y_m) = 0. \tag{25}$$

4. Whenever each of the systems S_1 and S_2 is made to reach equilibrium with a

third system S_3 under identical conditions, systems S_1 and S_2 are in mutual equilibrium.

The last condition is equivalent to the statement that of the three relations

$$
\begin{aligned}
F(x_0, x_1, \ldots, x_n;\ y_0, y_1, \ldots, y_m) &= 0, \\
G(x_0, x_1, \ldots, x_n;\ z_0, z_1, \ldots, z_k) &= 0, \\
H(y_0, y_1, \ldots, y_m;\ z_0, z_1, \ldots, z_k) &= 0,
\end{aligned}
\tag{26}
$$

which describe the equilibrium between S_1 and S_2, S_1 and S_3, and S_2 and S_3, in analogy with Eq. (25), any one is a necessary consequence of the other two.

This is possible only on condition that the system of Eqs. (26) is equivalent to a system of equations of the form

$$\rho(x_0, x_1, \ldots, x_n) = \sigma(y_0, y_1, \ldots, y_m) = \tau(z_0, z_1, \ldots, z_k).$$

In particular, in the circumstances, it is possible to replace relations (25) by the two equations

$$
\begin{aligned}
\rho(x_0, x_1, \ldots, x_n) &= \tau \\
\sigma(y_0, y_1, \ldots, y_m) &= \tau,
\end{aligned}
\tag{27}
$$

where τ is a new variable.

The variable τ is called *temperature*, and Eqs. (27) are equations of state for S_1 and S_2.

The system of equations (27) is, on the other hand, equivalent to the system

$$W\left\{\rho\right\} = \tau_1, \qquad W\left\{\sigma\right\} = \tau_1,$$

where W is an arbitrary function. It is seen that the functions (27) are not defined uniquely. This indeterminancy is expressed by the statement that the "temperature scale" can still be chosen arbitrarily.

Our assumptions lead to the further conclusion that at least one of the quantities $\partial\rho/\partial x_0$, $\partial\sigma/\partial y_0$ is not equal to zero identically. Indeed, if one of the quantities were to vanish, ρ and σ would depend only on $x_1, x_2, \ldots, x_n; y_1, y_2, \ldots, y_m$. However, these are all deformation coordinates that can be varied independently of each other; thus we could reach states for which it would be impossible to satisfy Eq. (25), and this contradicts our second assumption concerning thermal equilibrium.

One of these quantities, let us say $\partial\rho/\partial x_0$, is different from zero and it becomes possible, generally, to express x_0 as a function of the $(n + m + 1)$ remaining variables if use is made of the equation

$$\rho = \sigma.$$

For this reason it is possible to regard S as a system with $(n + m + 1)$ degrees of freedom and with $n + m$ deformation coordinates. According to our assumption on page 239, this is also a simple system and we may apply our considerations to it, too.

7. ABSOLUTE TEMPERATURE

According to the assumptions of the preceding section, the energy ϵ of our combined system S is equal to the sum of the energies ϵ_1 and ϵ_2 of its subsystems. By the same token, the external work A performed by S during an arbitrary process is equal to the sum of the quantities A_1 and A_2, which are associated with S_1 and S_2, respectively. Now, since the latter systems are simple ones, and since their coordinates have been normalized, we may write

$$d\epsilon_1 + DA_1 = M(x_0, x_1, \ldots, x_n)\, dx_0,$$

$$d\epsilon_2 + DA_2 = N(x_0, x_1, \ldots, x_n)\, dy_0;$$

adding the two equations, we obtain

$$d\epsilon + DA = M\, dx_0 + N\, dy_0. \tag{28}$$

Since system S is also a simple one, as demonstrated, we conclude, taking into account Eq. (27), that the right-hand side of Eq. (28) also possesses an integrating factor.

We now assume that there exists at least one system in nature whose equation of state contains one or more deformation coordinates. Experience teaches that this is indeed the case, for example, with gases. If we choose such a system for S_1, it follows that the function ρ contains at least one of the magnitudes $x_1, x_2, \ldots, x_n$, that is, for example, x_1. In this case, we see that M in Eq. (28) can be regarded as a function of

$$x_0, \tau, x_2, \ldots, x_n.$$

As far as the equation of state

$$\sigma(y_0, y_1, \ldots, y_m) = \tau$$

is concerned, we take into account all possible cases.

1. If σ does not depend on any variable of the set $y_0, y_1, \ldots, y_m$, we must insert a definitive value for τ in M. In the identity

$$du = \lambda(M\, dx_0 + N\, dy_0). \tag{29}$$

that has now arisen, the function u can depend only on x_0 and y_0 so that λM and λN also depend on these two variables only. Since now the y_i do not occur in M, λ cannot contain even a single deformation coordinate of S_2; however, the latter do not appear in λN either; it follows that N is also independent of these quantities and is a sole function of y_0.

2. In cases when σ depends on the single variable y_0, we would substitute for τ in M this function of y_0 and by the same reasoning as before we would conclude that N is solely a function of y_0.

3. Finally, we consider the case when σ contains at least one deformation variable. Let this be y_1; we can now regard this quantity, as well as N, as functions of

$$y_0, \tau, y_2, y_3, \ldots, y_m.$$

As far as Eq. (29) is concerned, λM and λN are still functions of x_0 and y_0 only. However, since M as well as λM does not contain $y_2, y_3, \ldots, y_m$, the same is true about λ. Furthermore, N is independent of these quantities, since otherwise λN would also contain these coordinates.

Employing an analogous reasoning with respect to the x_i it is possible to discover the result that λ depends at most on x_0, y_0, and τ, M at most on x_0 and τ, and N only on y_0 and τ.

Since, however, λM and λN are independent of τ as well, we find that

$$M\left(\frac{\partial\lambda}{\partial\tau}\right) + \lambda\left(\frac{\partial M}{\partial\tau}\right) = 0; \quad N\left(\frac{\partial\lambda}{\partial\tau}\right) + \lambda\left(\frac{\partial N}{\partial\tau}\right) = 0,$$

and conclude that the logarithmic derivative

$$\left(\frac{1}{\lambda}\right)\left(\frac{\partial\lambda}{\partial\tau}\right)$$

cannot depend either on x_0 or on y_0. Consequently, λ must be separable into a product of a function of the variable τ and a function of x_0 and y_0. Hence we may write

$$\lambda = \frac{\psi(x_0, y_0)}{f(\tau)},$$

from which it follows that M must be of the form

$$M = f(\tau)\alpha(x_0),$$

and for exactly the same reasons we must have

$$N = f(\tau)\beta(y_0).$$

The most general splitting of M and N into a product of factors one of which depends only on τ and the other only on x_0 or y_0 has the form

$$M = Cf(\tau)\left[\frac{\alpha(x_0)}{C}\right]; \quad N = Cf(\tau)\left[\frac{\beta(y_0)}{C}\right], \tag{30}$$

where C is an arbitrary constant different from zero.

Starting with a given temperature scale, we see that for a system whose equation of state contains one deformation coordinate, the factor $f(\tau)$ is fully determined except for a multiplicative constant. Moreover, such factors are the same for all such systems.

We note that the preceding decomposition of the function N is possible for systems that were examined under 1 and 2, and conclude that the function $f(\tau)$ possesses a very general physical meaning. The temperature scale defined by the function

$$t = Cf(\tau)$$

is called *absolute.* In order to fix the constant, it is customary to fix the difference between two fixed points, for example, between the steam point and the ice point of water at prescribed pressures.[9]

8. ENTROPY

In the case of every simple system, as seen from Eqs. (23) and (30), it is always possible so to normalize the coordinates as to render

$$DA = -\frac{\partial \epsilon}{\partial x_1} dx_1 - \frac{\partial \epsilon}{\partial x_2} dx_2 - \ldots - \frac{\partial \epsilon}{\partial x_n} dx_n$$

$$\frac{\partial \epsilon}{\partial x_0} = \frac{t\alpha(x_0)}{c},$$

where t is the absolute temperature. We now introduce a new coordinate η with the aid of the equation

$$\eta - \eta_0 = \int_{\alpha_0}^{x_0} \frac{\alpha(x_0)}{c} dx_0; \tag{31}$$

Eq. (31) can be solved for x_0, because M—and, according to Eq. (30), also α—is different from zero. Thus, the total differential of the energy function assumes the form

$$d\epsilon = t\, d\eta - DA. \tag{32}$$

Following Clausius, we call the new coordinate η "entropy"; it is defined by Eq. (31) except for an arbitrary, additive constant.

In relation to system $S = S_1 + S_2$, which was considered in the preceding section, we can write for every quasistatic process that

$$d\epsilon = d\epsilon_1 + d\epsilon_2 = t\,(d\eta_1 + d\eta_2) - DA_1 - DA_2,$$

[9] [*Editor's Note*: At present, the temperature of only one fixed thermometric point, the triple point of water, is given a conventional value (273.16 K).]

when η_1, η_2 denote the entropies of S_1 and S_2 respectively.

If we now express η_1 and η_2 with the aid of the equations

$$\eta_1 + \eta_2 = \eta$$

$$\frac{\partial \epsilon_1}{\partial \eta_1} - \frac{\partial \epsilon_2}{\partial \eta_2} = 0 \tag{33}$$

as functions of x_i, y_k, and the new variable η, we observe that the total differential for the total energy ϵ retains the form (32). This means that the entropy of the total system is equal to the sum of the entropies of its components. This additivity property of entropy that persists even if other partitions than those considered above (permeable to heat alone) are employed to separate S_1 and S_2 (as can be inferred from the theory of such partitions) has induced many physicists to regard entropy as a physical quantity that, like mass, fills every system of finite volume, and that, by contrast, depends on its instantaneous state.

According to Eq. (30), the entropy depends only on x_0 if normalized coordinates are used; it follows that the entropy remains constant in every quasistatic adiabatic process. Similarly, every process performed by a simple system during which the entropy remains constant is, according to our preceding argument, *reversible.*

9. IRREVERSIBLE PROCESSES

We have previously made the assumption regarding simple systems that during adiabatic processes all possible numerical values of work performed between a given initial state and a given final shape form a connected set of numbers. We now calculate the corresponding values of the final entropy η with the aid of the equation

$$\epsilon(\eta, x_1, x_2, \ldots, x_n) - \epsilon_0 + A = 0,$$

which is valid for *arbitrary* adiabatic processes; we shall discover that they also fill an entire segment. The initial value of entropy η_0 must necessarily constitute a point of this segment. This is because during quasistatic processes, which must also be counted among those taken into account at present, the value of η remains unchanged. If now η_0 were to be an *internal* point of this interval, it would become possible to impose by means of adiabatic processes, a value η placed in a certain neighborhood of η_0. It would further become possible arbitrarily to change the deformation coordinates at a constant value of η. This, however, conflicts with the axiom of the second law.

It follows that η_0 constitutes an *end-point* of the segment under consideration and that the value of the entropy during an arbitrary adiabatic process starting from a given initial state either never increases or never decreases.

If the initial state is now varied, the change in entropy for two arbitrary initial states of the same system must always proceed in the same direction. This is a consequence of continuity. Furthermore, the same is true for two different systems

S_1 and S_2 owing to the property of additivity of entropy, which was discussed in the preceding section.

Whether the entropy can only decrease or only increase depends also on the multiplicative constant C in Eq. (31) for entropy η. We choose this constant so as to make the absolute temperature t positive. Experience (which needs to be ascertained in relation to a single experiment only) then teaches that *entropy can never decrease.*

It follows from the preceding that equilibrium will always exist if all admissible virtual pocesses of the state of a simple system cause a decrease in entropy, and that it is impossible to devise an adiabatic process that would take the system under consideration from a given end state back to a given initial state if the entropy is not the same in both.

Every process during which the entropy varies is "irreversible."

10. ON THE POSSIBILITY OF EXPERIMENTALLY DETERMINING ENERGY, ENTROPY, AND ABSOLUTE TEMPERATURE

It is necessary to demonstrate that the quantities ϵ, η, and t, which we have introduced by our preceding arguments, can really be determined by experiment in every concrete case. The simplest way in which this can be done is to assume that it is possible to observe reversible processes with an adequate degree of accuracy. Since there are no logical reasons to prevent us from accepting this premise, we conclude that the science of thermodynamics can be justified by purely experimental means, that is, without adopting any assumptions concerning the nature of "heat."

In reality, if this method were adopted, we would encounter difficulties resulting from unavoidable experimental errors. For this reason, we shall divide the problem into two parts: *First,* we shall determine the absolute temperature scale; I am not in a position to enter into a discussion of this problem in the present paper because rather lengthy arguments are required for the purpose. However, if t is determined, it becomes relatively easy to calculate ϵ and η—as we shall see in this section—by an analysis of measurements that can be performed very accurately.

We assume that we are given a simple system S whose state depends on the variables

$$\xi_0, x_1, x_2, \ldots, x_n,$$

where x_i denote the deformation coordinates. Further, we assume that the following functions have been determined:

1. The equation of state of S in terms of an arbitrary temperature scale τ or

$$\psi(\xi_0, x_1, x_2, \ldots, x_n) = \tau. \tag{34}$$

2. The coefficients of a Pfaffian expression for the work A

$$DA = p_1\, dx_1 + p_2\, dx_2 + \ldots + p_n\, dx. \tag{35}$$

3. The coefficients of Pfaff's equation for quasistatic, adiabatic processes:

$$0 = \frac{1}{\partial\epsilon/\partial x_0}\,(d\epsilon + DA) = d\xi_0 + X_1\,dx_1 + X_2\,dx_2 + \ldots + X_n\,dx_n. \qquad (36)$$

In order to determine the coefficients $X_1, X_2, \ldots, X_n$, we picture the following experiments: We consider adiabatic processes of the system during which only one deformation coordinate, for example, x_1, increases by, say, Δx_1, whereas $x_2, x_3, \ldots, x_n$ remain constant; we now measure the change $\Delta\xi$ in ξ_0 during this process; if Δx_1 is small enough, and if the process is slow enough, we must have

$$x_1 = -\frac{\Delta\xi_0}{\Delta x_1}$$

for the given initial state.

According to our earlier results, the Pfaffian expression (36) must possess a multiplier λ that turns it into a perfect differential; this is only then possible when the quantities x_i satisfy certain differential equations that must exist for every particular system if our theory is correct.

Thus, we have the equation

$$\lambda\,(d\xi_0 + X_1\,dx_1 + X_2\,dx_2 + \ldots + X_n\,dx_n) = dx_0,$$

which can be integrated. Having chosen for x_0 one of its possible integrals, we introduce x_0 itself as a new coordinate; hence, from (34) we have

$$\phi(x_0, x_1, \ldots, x_n) = \tau, \qquad (37)$$

and Eq. (35) retains its original form, except that the p_i must be expressed as functions of the new variables.

As far as the absolute temperature is concerned, we choose the form

$$t = \beta(\tau),$$

and conclude from Eqs. (24) and (32) that

$$d\epsilon = \beta(\tau)\alpha(x_0)\,dx_0 - p_1\,dx_1 - \ldots - p_n\,dx_n. \qquad (38)$$

The functions α and β must be so determined as to render expression (38) integrable.

First, this is only possible if (35) is also integrable for a constant value of x_0. In turn, this imposes certain conditions on the p_i, which must be satisfied for every system. Further, we put

$$x_2 = a_2, x_3 = a_3, \ldots, x_n = a_n,$$

when the a_i denote constants, and introduce the notation

$$\tau = \phi(x_0, x_1, a_2, \ldots, a_n = f(x_0, x_1) \tag{39}$$

$$p_1(x_0, x_1, a_2, \ldots, a_n = g(x_0, x_1); \tag{40}$$

we can now see that the expression

$$\beta f(x_0, x_1)\alpha(x_0)\ dx_0 - g(x_0, x_1)\ dx_1$$

is a perfect differential, and this introduces the condition

$$\beta'\ [f(x_0, x_1)] \left(\frac{\partial f}{\partial x_1}\right) \alpha(x_0) + \frac{\partial g}{\partial x_0} = 0,$$

or

$$\beta'(\tau)\alpha(x_0) = -\frac{(\partial g/\partial x_0)}{(\partial f/\partial x_1)}. \tag{41}$$

The right-hand side of this equation must separate into a product of a function of x_0 and into one of τ, if we introduce into it x_1 as a function of x_0 and τ from Eq. (39). Thus, we may write

$$\frac{-(\partial g/\partial x_0)}{(\partial f/\partial x_1)} = \psi(x_0)\theta(\tau).$$

If now C and C' denote constants that remain to be determined, we obtain, finally,

$$\tau = \beta(\tau) = C\int_{\tau_1}^{\tau} \theta(\tau)\ d\tau + C',$$

$$\alpha(x_0) = \frac{\psi(x_0)}{C}.$$

The first of these equations represents the absolute temperature, whereas the second allows us to calculate the entropy.

$$\eta - \eta_0 = \frac{1}{C}\int_{\alpha_0}^{x_0} \psi(x_0)\ dx_0.$$

The constant of integration C is usually fixed by fixing by convention the difference D of the absolute temperature between two conventionally prescribed temperatures τ_1 and τ_2,[10] so that

$$C = D \bigg/ \int_{\tau_1}^{\tau_2} \theta(\tau)\ d\tau.$$

[10] [*Editor's Note*: See footnote on p. 247.]

Equation (38) can be integrated if the above values for α and β are introduced into it. This can always be done in concrete cases because the integrability conditions for them must be satisfied. We thus obtain

$$\epsilon - \epsilon_0 = C'(\eta - \eta_0) + F(\eta, x_1, \ldots, x_n).$$

It is seen that employment of reversible processes allows us to determine the absolute temperature except for an additive constant; the internal energy can be determined except for a linear function of entropy.

The preceding lack of definiteness can be obviated in that the constant C' can be calculated once for all with reference to a single *irreversible* process. For example, we can observe a process during which no external work is performed. In such cases the internal energy remains constant, whereas the state variables undergo measurable changes. This produces a linear equation for C'.

In the case of perfect gases, the calculation is as follows: The equation of state is

$$pv = \tau$$

if we choose the pressure p and volume v of a constant mass as the state variables. The equation of adiabatics is

$$pv^{\gamma+1} = \text{Const.}$$

Hence we may put

$$v = x_1 \quad \text{and} \quad pv^{\gamma+1} = x_0.$$

Equations (39) and (40) become

$$\tau = f(x_0, x_1) = x_0 x_1^{-\gamma}$$

$$p = g(x_0, x_1) = x_0 x_1^{-(\gamma-1)},$$

and Eq. (41) is now

$$\beta'(\tau)\alpha(x_0) = \frac{1}{\gamma}\, x_0,$$

so that

$$t = \beta(\tau) = C\tau + C', \quad \alpha(x_0) = \frac{1}{C\gamma}\, x_0,$$

$$d\epsilon = \left[\frac{x_1^{-\gamma}}{\gamma} + \frac{C'}{C_\gamma x_0}\right] dx_0 - x_0 x_1^{-(\gamma-1)}\, dx_1.$$

Integration of the last equation yields

$$\epsilon - \epsilon_0 = \frac{x_0 x_1^{-\gamma}}{\gamma} + C' \log \frac{x_0}{C\gamma},$$

or, reintroducing τ and ν,

$$\epsilon - \epsilon_0 = \frac{\tau}{\gamma} + \frac{C' l \tau}{C\gamma} + \frac{C' l \nu}{C}.$$

It is now necessary to invoke an irreversible process in order to determine C'. During an adiabatic expansion without work, τ remains constant. Hence $C' = 0$, and we can easily derive the well-known formulae

$$\epsilon = \frac{t}{C\gamma}, \qquad pv = \frac{t}{C}.$$

11. PRACTICAL DETERMINATION OF ϵ AND η

In this section we assume that the absolute temperature scale is known and denote the state variables of a simple system S by

$$\xi_0, x_1, x_2, \ldots, x_n$$

once again. We assume that we can measure the following data:

1. The equation of state.
2. The functions $p_1, p_2, \ldots, p_n$ in Pfaff's expression for the external work.
3. The "specific heat" at constant shape. The latter signifies that we can determine

$$\frac{\partial \epsilon}{\partial t}$$

at constant values of $x_1, x_2, \ldots, x_n$ either with the aid of calorimetric measurements or by observing certain irreversible processes.

It follows from item 1 that we can choose the $(n + 1)$ coordinates

$$t, x_1, x_2, \ldots, x_n$$

as our independent variables, where t again denotes the absolute temperature. Expressed in these coordinates, Eq. (32) assumes the form:

$$d\epsilon = t \left(\frac{\partial \eta}{\partial t} \right) dt + \sum_1^n \left(t \frac{\partial \eta}{\partial x_i} - p_i \right) dx_i.$$

Since this must be a perfect differential, the conditions of integrability give

$$\frac{\partial \eta}{\partial x_i} = \frac{\partial p_i}{\partial t} \qquad (i = 1, 2, \ldots, n). \tag{42}$$

In addition, taking into account item 3, we conclude that the quantity

$$\frac{\partial \eta}{\partial t} = \frac{(\partial \epsilon / \partial t)}{t} \tag{43}$$

is known. Equations (42) and (43) allow us to calculate the entropy η, except for an additive constant, on condition that certain integrability relations between the p_i and $\partial \epsilon / \partial t$ are satisfied. Substituting the values (42) into $d\epsilon$, we obtain

$$d\epsilon = t\left(\frac{\partial \eta}{\partial t}\right) dt + \sum_{1}^{n} \left[t\left(\frac{\partial p_i}{\partial t}\right) - p_i \right] dx_i, \tag{44}$$

which can be integrated as soon as the preceding conditions are satisfied.

Integrating Eqs. (42), (43), and (44) we obtain all required data.

We would be led to exactly analogous calculations if we measured the specific heats at constant values of the external forces instead of doing so for a constant shape. It would, however, be necessary to calculate first the "thermodynamic potential"

$$\epsilon - p_1 x_1 - p_2 x_2 - \ldots - p_n x_n$$

instead of the energy ϵ.

12. CRYSTALLINE SYSTEMS

All previous arguments and results can be extended to include the general case when some substances are solid and possess a crystalline nature.

The only difference is that the quantities that characterize such phases are different from those employed previously. In addition to the volume V it is necessary to introduce the deformation invariants as they have been defined in the theory of elasticity. Since we stipulate that the individual phases are homogeneous, these quantities assume the same value at all points of a phase, and can be employed as the thermodynamic coordinates for the whole phase.

Instead of the pressure, which ceases to be independent of direction, it is necessary to introduce the "mechanical invariants," that is, the coefficients that appear in the Pfaffian expression for external work and that constitute deformation coordinates. These, too, are known from elasticity theory.

In all, including the volume V, we need thirteen parameters instead of the two, namely V and p, which we have carried so far. In the case of special crystalline systems, this number can become reduced.

By contrast, the chemical coordinates $m_{\kappa i}$, which we studied in the first section of this paper, remain exactly the same as before.

Now, all these varied quantities are related to each other by certain equations, even if we consider single phases. This contrasts with the fact that previously we encountered additional equations of constraint that emanated only from contact surfaces. I again refer the reader to the theory of elasticity in the matter of these relations.

13. REMARKS ON THE RANGE OF VALIDITY OF THE LAWS OF THERMODYNAMICS

The method that we employed in order to derive the main results of thermodynamics (even though other ways for the establishment of this theory can be thought of), and, in particular, the concepts of "absolute temperature" and "entropy," compels us to suppose that these theorems and concepts are tied to many assumptions. Consequently, the range of validity of the preceding is correspondingly narrow.

Evidently, certain generalizations are immediately possible. For example, it is possible at once to drop the assumption that different substances must be homogeneous. To achieve this, it is merely necessary to regard the state variables, known to us from before, as functions of spatial coordinates. The definitions of energy and work performed must now be modified with the aid of appropriate integrals. The difficulties that arise as a result of such generalizations are of a purely mathematical nature; it is easy to remove them and to see that our results remain unaffected.

Similarly, it is easy to analyze the case when capillary forces must be included; a thermodynamic analysis of this problem can be found in the paper by Gibbs cited earlier; the latter appears to contain all principal ideas that are needed for a satisfactory solution of this problem.

Difficulties of an entirely different nature are introduced by problems of radiation, by problems of the "transfer of heat," and in particular, by the thermodynamics of substances in motion.

In the case of the simplest radiation phenomena, it is no longer possible to make do with a finite number of variables in order to define the state or the concept of equivalent systems. This is because the emissivity as well as the absorptivity and the dispersion of a substance must be defined for *every* wavelength so that here mere *numbers* no longer suffice. Instead, we must introduce *functions* that depend on one or more variables. The difference is the same as that encountered in mechanics when a transition is made from systems endowed with a finite number of degrees of freedom to the mechanics of continua.

We continue to realize that the concept of temperature is not a primary one. In other words, it should be possible to establish the various state variables without making use of this quantity. We have seen that the temperature enters our calculations through considerations concerning certain equilibrium conditions. We may now attempt to *define* the temperature of a radiating system S_1 with reference to the condition that it is in equilibrium with one of our earlier systems S_2 whose temperature is known to be t according to our previous considerations. It is, however, easy to see that such a definition can lead to contradictions: This is because system S_2 must also be regarded as a radiating body, and it is feasible to think that there exist two systems, S_2' and S_2'', which are in identical state from the point of view of ordinary thermodynamics employing fewer state variables, but differ from each other when examined from the point of view of radiation. For this reason, S_1 need not be in simultaneous equilibrium with S_2' and S_2'', and whenever this is the case it is impossible to assign a temperature t in the ordinary sense of the term. The preceding question must be subjected to a detailed analysis before an answer to it can be given. Similar reservations apply to the concept of entropy because its definition is so closely related to that of absolute temperature.

Insofar as the thermodynamics of fluids in motion is concerned (here one can include also the theory of heat flow), the difficulties are of a different nature as long as the processes can be analyzed without regard to radiation.

In this case we could, presumably, assign a certain temperature to every point of the system and to regard it as a function of time. It is, therefore, possible that such a temperature would depend on all state variables, that is, in this case, also on the velocities.

In spite of all this, we could not employ our earlier methods for the determination of the energy function, because all processes have now become irreversible owing to internal friction that must not be neglected in them.

13

Reprinted with permission from pp. 50–83 of *Introduction to Theoretical Physics, Vol. V: Theory of Heat*, Henry L. Brose, trans., Macmillan and Co. Ltd., London, 1932, 310 pp.

SECOND LAW OF THERMODYNAMICS

M. Planck

§ 36. THE content of the second law of thermodynamics is sharply distinguished from that of the first law because it concerns a question which is not touched on by the latter at all, namely the question of the direction in which a thermal process occurs in nature. For not every change which is compatible with the principle of conservation of energy satisfies the further condition imposed by the second law on the processes that actually occur in nature. If, for example, an exchange of heat occurs by conduction between two bodies at different temperatures, the first law requires only that the amount of heat given out by the one body should equal the amount of heat taken up by the other body. Whether the heat conduction occurs in the one or the other direction cannot be decided on the strength of the first law. Indeed, the concept of temperature is in itself foreign to the energy principle, as can be seen from the fact that this principle does not lead to an exact definition of temperature.

As for the direction in which processes occur in nature and the way in which this question is answered by the second law there is an essential difference between mechanical and electrodynamic events on the one hand and thermo-chemical events on the other hand—a difference to which we have already alluded in § 2. For whereas the former can always also occur in the exactly opposite direction—a heavy body can rise just as well as it can fall, a spherical electrodynamic wave can propagate itself just as well inwards as outwards—according to the second law no thermal event can be directly reversed.

The problem of formulating the second law correctly has occupied physicists for decades. A long time passed before it was recognized that the content of the second law is not exhausted if—as was done occasionally even by Clausius and later with renewed emphasis by Ostwald—every process in nature is resolved into a series of energy transformations and the direction of each individual transformation is enquired into. It is true that in each individual case we can name the different kinds of energy that are transformed into one another—this follows from the first law—but there always remains a certain arbitrariness as to how the individual transformations are allocated to one another, and this arbitrariness cannot be removed by a general convention.

Even nowadays the nature of the second law is sometimes sought in the tendency of natural phenomena to "degrade" energy on the ground that, for example, mechanical energy can be completely transformed into heat but heat can be transformed only incompletely into mechanical energy, in the sense that if a quantity of heat is transformed into mechanical energy then another transformation, such as a thermal transition from a higher to a lower temperature which serves as a compensation process, must always occur simultaneously.

That this formulation, which is useful in special cases, by no means gets to the root of the matter can be seen from the following simple illustration. If a gas is allowed to expand, doing work in the process, and if at the same time the temperature of the gas is kept constant by the transference of heat from a reservoir at a higher temperature, we may say that the heat transferred by the reservoir has been *completely* transformed into work. For while the gas has retained its temperature it has also retained its internal energy unchanged (§ 24), and other transformations of energy are not occurring. No fact of any kind can be objected to in this assertion. But in the case of the second law we are concerned with particular facts that can be ascertained by measurement. That is also why the second law cannot be deduced *a priori*. We can speak of proving it only in so far as its total content may be deduced from a single simple fact of experience of convincing certainty.

§ 37. In connexion with what has just been said we shall now base the general proof of the second law on the following empirical law: "it is impossible to construct a machine which functions with a regular period and which does nothing but raise a weight and cause a corresponding cooling of a heat reservoir."

Such a machine could be used simultaneously as a motor and as a cooling machine without any other expenditure of energy or materials. It would at any rate be the most advantageous in the world. It would not be equal to the perpetual motion machine, for it by no means produces work from nothing, but from the heat which it abstracts from the reservoir. That is also the reason why it does not, like the perpetual motion machine, conflict with the first law. But it would nevertheless possess the most essential advantage of the perpetual motion machine for mankind, that of supplying work without expenditure. For the heat contained in the soil, in the atmosphere and in the ocean is always available, just like the oxygen of the air, in inexhaustible quantities for direct use by anyone. This circumstance also accounts for our beginning with the above empirical law. For as we shall deduce the second law from it we secure for ourselves, in the event of our ever discovering any deviation of a natural phenomenon from the second law, the prospect of immediately being able to apply it practically in a very important way. For as soon as any phenomenon is found that contradicts a single inference from the second law, the contradiction would be due to an inaccuracy in the very first assumption on which it is based and it would be possible, by following the above reasoning backwards step by step, to use the phenomenon to construct the machine above mentioned. For the sake of brevity we shall in the sequel follow a suggestion of Ostwald and call it a " perpetual motion machine of the second kind," since it bears the same relation to the second law that the perpetual motion machine of the first kind bears to the first law.

§ 38. If we compare the two kinds of perpetual motion machine we at once observe a fundamental difference : the law forbidding the perpetual motion machine of the first kind also applies conversely, that is, work can neither be produced absolutely nor annihilated absolutely, whereas the law which forbids the perpetual motion machine of the second kind does not apply conversely; that is, it is certainly possible to construct a machine which does nothing more than lower a weight while a heat reservoir is correspondingly warmed. As an illustration of this kind of machine we have that used by Joule to measure the mechanical equivalent of heat; it is set into motion by means of a falling weight which causes rotating paddle-wheels to warm a liquid by friction. For if the weight reaches the floor with vanishingly small velocity no change has taken place in nature except that the liquid—which is here to be regarded as a heat reservoir —has been warmed. In fact it is clear that every frictional process represents a reversed perpetual motion machine of the second kind, so that our empirical law stated above may also be formulated as follows : there is no possible way of completely reversing a process in which heat is generated by friction. The word " completely " is used here to express that the initial state of the frictional process has everywhere been exactly restored. To take a definite example, if after a Joule's friction experiment had been carried out it were possible by some process to bring the fallen weight back to its original height and to cool the liquid correspondingly without any other changes remaining this would obviously be a perpetual motion machine of the second kind. For it does nothing beyond raising a weight and correspondingly cooling a heat reservoir.

For the sake of brevity we shall call a process which can

in no way whatsoever be completely reversed "irreversible" and all other processes "reversible." For a process to be reversible, then, it is not sufficient to restore the bodies that participate in the process to their initial state—this is always possible in principle—but it is required that it should in some way be possible to restore the initial condition of the process everywhere in nature, no matter what technical devices and mechanical, thermal, chemical and electrical apparatus are used. All that is essential is that any material and apparatus used should at the end be again in exactly the same state as in the beginning when they were taken for use.

§ 39. Any process that occurs in nature is either reversible or irreversible. We have as examples of reversible processes all purely mechanical and electrodynamic processes. For if they occur in the reverse direction the initial state is completely restored. As an example of an irreversible process we have had the generation of heat by friction; other examples will be given in the sequel.

The significance of the second law consists in the fact that it furnishes us with a necessary and sufficient criterion as to whether a definite process that occurs in nature is reversible or irreversible. Since the decision on this question depends only on whether the process can be completely reversed or not, we are concerned only with the constitution of the initial state and the final state of the process but not on its intermediate course. For it is merely a question as to whether, starting from the final state, we can or cannot again arrive at the initial state without anything being changed. Hence the second law furnishes for any process whatsoever in nature a relationship between those quantities which refer to the initial state and those which refer to the final state. In the case of irreversible processes the final state is distinguished by a certain property from the initial state, whereas in the case of reversible processes these two states are in a certain sense of equal value. We shall express this briefly by saying that in the case of an irreversible process the final state has a greater " thermodynamic probability " or a greater " thermodynamic weight " (we say " thermodynamic " to distinguish it from " mathematical " probability or mechanical weight, respectively) than the initial state. These words are to convey no more than what has been said above. We may then formulate the content of the second law by saying that it gives us a measure of the magnitude of the thermodynamic probability or the thermodynamic weight of a physical configuration in a given state. Our next task is to find this measure.

§ 40. To solve this problem we shall adopt the following course. Let us imagine an arbitrary system of bodies and let us consider any two different and exactly defined states of the system, which we shall denote by Z and Z'.

The question is whether and under what conditions a process in nature is possible such that it transfers the system of bodies in some way from the state Z to the state Z' or conversely, without anything being left changed outside the system. We can make the last proviso superfluous by including all the bodies in the world in the system under consideration. In other words, our object is to specify to which of the two states Z and Z' the greater thermodynamic probability is to be assigned. There are clearly three different possibilities. A transition may be possible both from the state Z to the state Z' and also conversely from Z' to Z; this process is then reversible every time and, indeed, in all its parts, and so the probabilities are equally great for both states. Or a transition may be possible from Z to Z' but not reversely from Z' to Z; the process in question is then irreversible and the probability of Z' is greater than that of Z. Or, lastly, the opposite case may occur, namely if an irreversible process from Z' to Z is possible of execution, so that Z has a greater probability than Z'.

If the definition of the two states Z and Z' is quite arbitrary at the outset, they must nevertheless fulfil the condition that the transition from one state to the other involves the loss neither of matter nor of energy. For otherwise the process in question would be impossible from the very start. The system of bodies must therefore have the same chemical constituents and the same energy in both states. Otherwise, however, Z and Z' may be selected arbitrarily.

§ 41. We start from the simplest case that the states Z and Z' are distinguished from each other only by the behaviour of a single homogeneous body of the kind considered above in § 22. Let this body have the volume V and the temperature θ in the state Z, the temperature being measured by any thermometer, for example, a mercury thermometer or any gas thermometer; and in the state Z' let it have the volume V' and the temperature θ'. The volume and the temperature also determine the energies of the body, U and U', respectively. Since U and U' are in general different from each other and since on the other hand we must ensure that the transition from Z to Z' satisfies the energy principle, we imagine an invariable weight G included in the configuration in question, the centre of gravity of this weight being at a height h in the state Z and at a height h' in the state Z', so that:

$$G.(h - h') = U' - U \quad . \quad . \quad . \quad . \quad (70)$$

From the standpoint of the energy principle a transition of the system from one state to the other is possible without anything remaining changed outside the system.

We now start from the state Z and endeavour to reach the state Z'. To do this we first subject the body in question to a reversible adiabatic change, as was done in § 32 with a gas. We then have $Q = 0$ and by (49):

$$dU + pdV = 0 \quad . \quad . \quad . \quad . \quad . \quad (71)$$

Also:

$$dU = \left(\frac{\partial U}{\partial \theta}\right)_V d\theta + \left(\frac{\partial U}{\partial V}\right)_\theta dV$$

Thus we get:

$$\left(\frac{\partial U}{\partial \theta}\right)_V d\theta + \left\{\left(\frac{\partial U}{\partial V}\right)_\theta + p\right\} dV = 0$$

The expression on the left-hand side is not a complete differential, as we saw in § 27. But there is always an "integrating factor" N, that is, a function of the two independent variables θ and V which when divided into the expression makes it a complete differential, so that we may always write:

$$\frac{dU + pdV}{N} = dS \quad . \quad . \quad . \quad . \quad (72)$$

where S is now a finite function of the two independent variables θ and V. We may thus regard S, just like the energy U, as a definite property of the state of the body. We shall follow Clausius and call it the "entropy" of the body in the state defined by θ and V.

But the definition of the entropy given by (72) is not yet unique. For there is not only one, but, indeed, an infinite number of quantities N, which when divided into the expression (71) make it a complete differential. This is easily seen by writing $N.f(S)$ instead of N, where f denotes any arbitrary function of a single variable. Corresponding to every expression for f there is then a different complete differential (72) and hence a different definition for the entropy.

Hence there is in the quantity N one factor, dependent only on S, which can be selected arbitrarily; and to complete the definition of entropy it is necessary to fix this factor. For the present, however, we shall leave our decision on this question open and shall calculate for the present with any arbitrarily chosen N, which we take as positive:

$$N > 0 \quad . \quad . \quad . \quad . \quad . \quad . \quad . \quad (73)$$

The following theorems hold independently of the arbitrariness still left in the definition of entropy.

Now that S has been determined by (72) we may integrate the differential equation (71) and we obtain:

$$S = \text{const.} \quad . \quad . \quad . \quad . \quad . \quad . \quad (74)$$

a definite relation between θ and V which holds for the process described, or a relation between U and V, the so-called reversible adiabatic relation.

§ 42. To pass from the state Z to the state Z' let us now make V transform into V', U transforming into U' during the process, by (74). At the same time a certain amount of mechanical work is performed by or on the system; let it correspond to the transference of the weight G from the height h to the height h^*. We have, by the energy principle:

$$G . (h^* - h) = U - U^*$$

or, by (70):

$$G . (h^* - h') = U' - U^* \quad . \quad . \quad . \quad (75)$$

Three cases are then possible:

1. $U^* = U'$. The body then satisfies the conditions of the state Z', and since the weight G is then also at the height $h^* = h'$ determined by (75), the state Z' is completely attained; the desired transition is then realizable, the process being reversible. Thus in this case the states Z and Z' have the same probability.

2. $U^* < U'$. The transferred energy of the body, U^*, is then less than in the state Z'. In this case the state Z' may be attained by heating the body by friction, the volume V' being kept constant; this is done by allowing the weight G which, by (75), is situated at the height $h^* > h'$, to sink to the level h'. The state Z' is then completely realized but now, according to § 39, by an irreversible process, that is, the state Z' possesses a greater probability than the state Z.

3. $U^* > U'$. Then $h^* < h'$ and the transition to the state Z' is impossible as it would represent a perpetual motion machine of the second kind (§ 38). Hence in this case the state Z has a greater probability than the state Z'.

Let us next enquire into the behaviour of the entropy of the body in question in the three different cases. By the definition of entropy (72) and by (73) the entropy changes in the same sense as the energy, if the volume remains constant ($dV = 0$). Now when the volume V' and the energy U^* are attained the entropy has retained its original value S by (74), the difference of S and S', that is, the difference of the entropies of the states (V', U^*) and (V', U') has the same sign as the difference of U^* and U'. Thus in the first case $S = S'$; in the second $S < S'$; in the third $S > S'$.

From this it also follows conversely that according as the entropy S' of the body in the state Z' is just as great, greater than or less than the entropy S of the body in the state Z, the state Z' has a probability which is just as great as, greater than or smaller than that of the state Z.

§ 43. The last deduction leads us directly to a theorem of far-reaching importance. If any arbitrary physical configuration has passed by means of some reversible physical or chemical process from a state Z to another state Z', which differs from the state Z only by the circumstance that a single body of the kind just considered has undergone a change and that a weight has correspondingly shifted its centre of gravity, then the entropy of the body in both states is the same. For if it had become greater the transition from Z to Z' would be irreversible according to the preceding paragraph, which would contradict the initial assumption. And if it had become smaller the transition would be impossible, which would also lead to a contradiction. But if the supposed process was irreversible, then the entropy of the body in the state Z' is necessarily greater than in the state Z.

A simple illustration is given by the adiabatic expansion of a gas without the performance of external work, which was described in §§ 23 and 24. Since, for this, $dU = 0$ and $dV > 0$, we have by (72) that $dS > 0$; that is, this process is irreversible, just like friction.

§ 44. We shall now assume that the two states Z and Z' given from the outset differ owing to the different behaviour of *two* bodies, which we shall denote by 1 and 2. Let them be characterized in the state Z by the values θ_1, V_1, θ_2, V_2 and in the state Z' by the values θ'_1, V'_1, θ'_2, V'_2. Then the internal energies and the entropies of the bodies in the two states are determined, the entropies being fixed except for the arbitrary factor still involved in the definition of entropy.

To make a transition from Z to Z' possible at all we shall suppose an invariable weight G to be included in the system of bodies, the centre of gravity of G in the state Z being at the level h and in the state Z' at the level h' so that:

$$G(h - h') = (U'_1 + U'_2) - (U_1 + U_2) \quad . \quad . \quad (76)$$

The mechanical work necessary for the transition from Z to Z' is then available to exactly the right amount.

Starting from the state Z we now again endeavour to effect the transition to the state Z' by means of an irreversible process. So long as we treat each individual body adiabatically their original entropies S_1 and S_2 remain constant by (74) and we make no progress. But we have a means of altering the entropies in a reversible way. For we first bring the two bodies singly by an adiabatic reversible path to a quite arbitrary temperature θ and then put them into thermal connexion with each other (but not so that their pressures can equalize). This does not disturb the thermodynamic equilibrium, and the two bodies now represent a single system capable of certain reversible changes, and its state is determined by three mutually independent variables θ, V_1, V_2.

If we now subject this composite system to a further reversible adiabatic process by slowly altering the volumes V_1 and V_2 independently of each other in some way by compression or dilatation, the change in the total internal energy is, by the first law, equal to the total external work, thus:

$$dU_1 + dU_2 + p_1 dV_1 + p_2 dV_2 = 0$$

or, by (72):

$$N_1 dS_1 + N_2 dS_2 = 0 \quad . \quad . \quad . \quad . \quad (77)$$

In this algebraic sum the first summand denotes, by (49), the heat transferred to the first body from without while the second term denotes the heat transferred to the second body, which is equal and opposite to the former amount. Equation (77) imposes a condition on the three variables θ, V_1 and V_2, so that only two of them, say θ and V_1, may be arbitrarily chosen, while the third, V_2, is then fixed. Hence by a reversible process of this kind it is possible to bring the body 1 into any arbitrary state, whereas the state of the body 2 has necessarily to adjust itself to that of the body 1.

§ 45. But we can assert still more. Every time when the body 1 assumes its original entropy S_1 at an arbitrary temperature θ the second body 2 also possesses its original entropy S_2. For as soon as the entropy of the body 1 has again become S_1 the bodies can be separated and the body 1 can be brought alone into its initial state (θ_1, V_1) by means of a reversible adiabatic process. The state of the system of bodies which is produced in this way then differs from the original state Z only in the behaviour of the body 2; and since the whole process is reversible, the entropy of this body is, by § 43, the same as at the beginning, namely, S_2. And, indeed, corresponding to a definite value of the entropy of the body 1 there is always a perfectly definite value of the entropy of the body 2. Otherwise the general theorem of § 43 would be contradicted.

In other words, if in place of the independent variables θ, V_1, V_2 we introduce the independent variables θ, S_1 and S_2 in the equation (77), θ disappears from the equation altogether and it reduces itself to a relation of the form:

$$F(S_1, S_2) = 0$$

or, expressed in differential form:

$$\frac{\partial F}{\partial S_1} dS_1 + \frac{\partial F}{\partial S_2} dS_2 = 0 \quad . \quad . \quad . \quad . \quad (78)$$

But in order that (77) should merge generally into (78) it is necessary and sufficient for the differential expressions of the two equations to differ only by a factor of proportionality:

$$N_1 dS_1 + N_2 dS_2 = N\left(\frac{\partial F}{\partial S_1} dS_1 + \frac{\partial F}{\partial S_2} dS_2\right)$$

or:

$$N_1 = N\frac{\partial F}{\partial S_1} \text{ and } N_2 = N\frac{\partial F}{\partial S_2}$$

Consequently:

$$\frac{N_1}{N_2} = \frac{\dfrac{\partial F}{\partial S_1}}{\dfrac{\partial F}{\partial S_2}}$$

That is, the quotient of N_1 and N_2 depends only on S_1 and S_2 but not on θ. But since N_1 and N_2 are themselves determined by θ and S_2 they are necessarily of the form:

$$N_1 = f_1(S_1) \,.\, T \text{ and } N_2 = f_2(S_2) \,.\, T \quad . \quad . \quad (79)$$

respectively, where f_1 and f_2 are functions of a single argument and the function T which is the same for both bodies depends only on the temperature θ. The quantity T contains a constant arbitrary factor. We choose its value positive and fix its unit by making the difference of the values which T assumes for $\theta = 100$ and $\theta = 0$ equal to 100 thus:

$$T_{100} - T_0 = 100 \quad . \quad . \quad . \quad . \quad (80)$$

and we call the quantity T which is completely defined in this way the "absolute temperature" of the two bodies. By (72) *the absolute temperature of a body is the positive temperature function which satisfies the normalizing equation* (80) *and which when divided into the incomplete differential* $dU + pdV$ *converts it into a complete differential.* Concerning the way in which it is measured see § 49.

To complete the definition of entropy as well we bear in mind that the functions f_1 and f_2 in (79) are positive on account of the conventions about the signs of N_1, N_2 and T, but otherwise, as is clear from the discussion of § 41, they may be arbitrarily chosen. We therefore set $f_1 = 1$, $f_2 = 2$ and thus obtain from (79):

$$N_1 = N_2 = T \quad . \quad . \quad . \quad . \quad . \quad (81)$$

And from (72) we get:

$$\frac{dU + pdV}{T} = dS \quad . \quad . \quad . \quad . \quad (82)$$

or, referred to unit mass:

$$\frac{du + pdv}{T} = ds \; . \quad . \quad . \quad . \quad . \quad (83)$$

is obtained as the entropy of a homogeneous body whose state is determined by its temperature and its volume. In the expression for the entropy there is still, as we see, an arbitrary additive constant.

§ 46. The results just obtained make it easy for us to give a complete answer to the question proposed in § 44 about the conditions governing the transition of the two bodies in question from the state Z to the state Z'. For in the reversible process described, to which the two thermally coupled bodies are subjected, we have by (77) and (81):

$$dS_1 + dS_2 = 0$$

Thus:

$$S_1 + S_2 = \text{const.} \quad . \quad . \quad . \quad . \quad (83a)$$

That is, the sum of the entropies remains constant. So if we bring the body 1 to the desired state of entropy S'_1 the body 2 acquires the entropy:

$$S^*_2 = S_1 + S_2 - S'_1 \quad . \quad . \quad . \quad . \quad (84)$$

If we then separate the two bodies and bring the volume of the first to V'_1 and its temperature to θ'_1 by means of a reversible adiabatic process, then the first body is in the desired final state and may be left out of consideration in the sequel. All that we are concerned with now is whether S^*_2 is as great, smaller than or greater than S'_2 or, what amounts to the same, by (84), whether:

$$S_1 + S_2 \overline{\gtrless} S'_1 + S'_2$$

In the first case the state Z' is fully attainable according to § 42, namely, by a reversible process; in the second case the state Z' is attainable by an irreversible process; in the third case the state Z' cannot be reached at all.

Hence the sum of the entropies of the two bodies in any state is a measure of the probability of this state.

§ 47. It now remains to generalize the last theorem for any arbitrary number of bodies. Let us imagine a system of n such bodies in two quite arbitrary states Z and Z' and enquire what is the condition that a transition from Z to Z' should be possible without changes of any sort remaining in other bodies.

In order that the transition should satisfy the energy principle we include an invariable weight in the system, whose centre of gravity in the state Z is at the level h and in the state Z' at the level h', so that:

$$G.(h-h')=(U'_1+U'_2+\ldots+U'_n)-(U_1+U_2+\ldots+U_n)$$

Starting from the state Z we now put the bodies 1 and 2 into thermal communication with each other in the manner described in § 44 and so bring the body 1 by a reversible process into the state of entropy S'_1. We then separate the two bodies and proceed in just the same way with the bodies 2 and 3 by bringing the body 2 to the state of entropy S'_2. Proceeding in the same way we bring the body $n-1$ to the state of entropy S'_{n-1}. Let the body n then have the entropy S^*_n. Since during each one of these reversible processes the total sum of all the entropies of all the bodies must on account of (83a) remain constant, we have:

$$S^*_n=(S_1+S_2+\ldots+S_n)-(S'_1+S'_2+\ldots+S'_{n-1}) \quad (85)$$

Now each of the bodies from 1 to $n-1$ may separately be brought along a reversible adiabatic path into its desired final state. Hence the decision as to whether the state Z' is fully attainable depends on whether the entropy S^*_n of the body n is just as great, smaller than or greater than S'_n or, by (85), if we denote the sum of the entropies of all the n bodies by ΣS, whether ΣS is just as great as, smaller than or greater than $\Sigma S'$. In the first case the state Z' is attainable by a reversible process, in the second by an irreversible process, whereas in the third the transition is altogether impossible.

From this it follows conversely that *every physical or chemical process that occurs in nature takes place in such a way that the sum of the entropies of all the bodies that undergo change in the process either remains unchanged or becomes augmented.* Expressed more briefly: *the entropy is a measure of the thermodynamic probability* (§ 39).

In a reversible process the sum of the entropies remains unchanged. But the reversible processes in reality are only ideal limiting cases which are, however, of considerable importance for theoretical considerations.

The content of the second law of thermodynamics may be regarded as exhaustively described in the above sentences.

§ 48. Passing on to the applications of the second law we shall next investigate the conclusions that follow from the fact that the expression (83) for the entropy of unit mass of a homogeneous body is a complete differential. If we choose T and v as independent variables, we have:

$$du = \left(\frac{\partial u}{\partial T}\right)_v dT + \left(\frac{\partial u}{\partial v}\right)_T dv$$

and by (83):

$$ds = \frac{1}{T}\left(\frac{\partial u}{\partial T}\right)_v dT + \frac{1}{T}\left\{\left(\frac{\partial u}{\partial v}\right)_T + p\right\} dv$$

On the other hand:

$$ds = \left(\frac{\partial s}{\partial T}\right)_v dT + \left(\frac{\partial s}{\partial v}\right)_T dv$$

Consequently, since dT and dv are independent of each other:

$$\left(\frac{\partial s}{\partial T}\right)_v = \frac{1}{T}\left(\frac{\partial u}{\partial T}\right)_v \quad . \quad . \quad . \quad . \quad . \quad (86)$$

and:

$$\left(\frac{\partial s}{\partial v}\right)_T = \frac{1}{T} \cdot \left\{\left(\frac{\partial u}{\partial v}\right)_T + p\right\} \quad . \quad . \quad . \quad . \quad (87)$$

F

If we differentiate the first equation with respect to v, and the second with respect to T we get, by equating the two expressions so obtained:

$$\left(\frac{\partial u}{\partial v}\right)_T = T\left(\frac{\partial p}{\partial T}\right)_v - p \quad . \quad . \quad . \quad . \quad (88)$$

and by (52), (86) and (87):

$$\left(\frac{\partial s}{\partial T}\right)_v = \frac{c_v}{T} \quad . \quad . \quad . \quad . \quad . \quad . \quad . \quad (89)$$

$$\left(\frac{\partial s}{\partial v}\right)_T = \left(\frac{\partial p}{\partial T}\right)_v \quad . \quad . \quad . \quad . \quad . \quad . \quad (90)$$

where c_v now denotes the specific heat referred to the absolute temperature.

By differentiating (89) with respect to v and (90) with respect to T we obtain the relation:

$$\left(\frac{\partial c_v}{\partial v}\right)_T = T\left(\frac{\partial^2 p}{\partial T^2}\right)_v \quad . \quad . \quad . \quad . \quad . \quad . \quad (91)$$

which brings the connexion between the specific heat and volume into relationship with the connexion between the thermal coefficient of expansion (§ 8) and the temperature. Both quantities are very small in the case of gases.

§ 49. The relations just derived can now serve as a measure of the way in which the absolute temperature depends on any conventional temperature. So far we have had to remain satisfied with a conventional temperature θ referred to an arbitrarily chosen thermometric substance (see § 4), and it was only on grounds of expediency that in § 5 we decided in favour of using a gas as the thermometric substance. But the uniformity achieved in this way was only of a practical nature and involved no matter of principle. For, strictly speaking, all gases behave differently, particularly at high densities and low temperatures, and therefore, to be accurate, we should select some definite gas as a thermometric gas. Hence it is of the greatest importance to have a process of measurement which enables us quite generally to reduce the conventional temperature θ of a body to its absolute temperature T. In principle we can adapt *every* equation which follows from the second law for measuring T (cf. § 61 below). Let us take, for example, the equation (88) and introduce the independent variables θ and v into it in place of T and v. For it is these variables θ and v with which all measurements are effected. Since T depends only on θ, we obtain :

$$\left(\frac{\partial u}{\partial v}\right)_{\theta} = T\left(\frac{\partial p}{\partial \theta}\right)_{v} \cdot \frac{d\theta}{dT} - p$$

Here $\left(\frac{\partial u}{\partial v}\right)_{\theta}$, p and $\left(\frac{\partial p}{\partial \theta}\right)_{v}$ are to be regarded as measurable functions of θ and v. This differential equation may be integrated in the following way :

$$\int_{0}^{\theta} \frac{\left(\frac{\partial p}{\partial \theta}\right)_{v} d\theta}{\left(\frac{\partial u}{\partial v}\right)_{\theta} + p} = \int_{T_0}^{T} \frac{dT}{T} = \log T - \log T_0 \quad . \quad . \quad (92)$$

and hence :

$$\log T_{100} - \log T_0 = \int_{0}^{100} \frac{\left(\frac{\partial p}{\partial \theta}\right)_{v} d\theta}{\left(\frac{\partial u}{\partial v}\right)_{\theta} + p} \quad . \quad . \quad (93)$$

The last two equations together with the normalizing equation (80) determine T completely as a function of θ, and so the conventional temperature is reduced to the absolute temperature.

In particular the values of T_{100} and T_0 may be derived from (93) and (80).

In the integration on the right-hand side the volume v must clearly disappear altogether; this requirement can also be used to test the second law. The numerator of the integrand is obtained directly from the equation of state of the body, but the denominator is derived from the quantity of heat which the body takes up from without during reversible isothermal expansion or, respectively, gives up to its surroundings during reversible isothermal compression. For by equation (50) of the first law the ratio of the transferred quantity of heat q to the change of volume dv during a reversible isothermal expansion is :

$$\left(\frac{q}{dv}\right)_0 = \left(\frac{\partial u}{\partial v}\right)_0 + p$$

§ 50. We shall now assume in particular that θ is measured by a gas thermometer and we choose as our thermometric substance hydrogen, as in § 5. Then the coefficient of expansion α that occurs in the equation of state (9) of the gases is constant only for hydrogen; for all other gases it is different and varies with the temperature θ. Thus if we apply the last equations to any gas very simple expressions present themselves for the pressure p and the energy u, but all the simple relations that we have deduced above for gases have the disadvantage that they are qualified by the word "nearly." Not a single one of them holds rigorously. To escape from this unsatisfactory state of affairs we proceed to introduce by definition a certain type of gases, which we call "ideal gases" and which satisfy the above simple relations completely. Thus we define as the equation of state of an ideal gas the equation (9), and as the energy of an ideal gas the expression (56). We now determine the relation between the absolute temperature T and the conventional temperature θ referred to an ideal gas as the thermometric substance. By substituting the values of p and u in (92) we get after integration :

$$\log T - \log T_0 = \log (1 + \alpha\theta)$$

Thus :

$$T = T_0 + \alpha T_0 \theta \quad . \quad . \quad . \quad . \quad . \quad (94)$$

Likewise, by (93) :

$$T_{100} = T_0 + 100\alpha T_0.$$

From this it follows by (80) that:

$$\alpha T_0 = 1.$$

That is, the coefficient of expansion of an ideal gas is the reciprocal of the value of the absolute temperature of the freezing point of water and hence has *the same value for all ideal gases.* Since the numerical value of α is sufficiently well known we have for the absolute temperature of the freezing point of water, by (6) :

$$T_0 = \frac{1}{\alpha} = 273{\cdot}2$$

and, in general, by (94) :

$$T = 273{\cdot}2 + \theta \quad . \quad . \quad . \quad . \quad . \quad (95)$$

Thus the absolute temperature is nothing more than the conventional temperature referred to an ideal gas as the thermometric substance but with the zero point displaced. Since, by definition, T is positive, we also have $\theta > -273$. The limiting point $T = 0$ is called the absolute zero of temperature. It is not attainable practically because the integral in (92) becomes infinite for it.

In future we shall as a general rule use the absolute temperature in our calculations. The equation of state of any ideal gas or of a mixture of ideal gases is then, by (35) :

$$p = \frac{RnT}{V} \quad . \quad . \quad . \quad . \quad . \quad . \quad (96)$$

where n denotes the total number of moles and R is the numerical factor which has the value (55) and which is now equally great for all ideal gases and is hence called the "absolute gas constant." Further the energy of an ideal gas is, by (56) :

$$u = c_v T + \text{const.} \quad . \quad . \quad . \quad . \quad . \quad (97)$$

Although the different ideal gases satisfy the same equation of state (96) they nevertheless have different specific heats and different molar heats between which, however, the relation (55) of course always holds.

§ 51. The equation (88) together with the equation (54) of the first law in which, as always, by (95), $d\theta$

can be put in place of dT, leads to the generally valid relation :

$$c_p - c_v = T\left(\frac{\partial p}{\partial T}\right)_v \cdot \left(\frac{\partial v}{\partial T}\right)_p \quad . \quad . \quad . \quad (98)$$

which may be applied to calculating c_v from c_p. Since it is not easy to measure $\left(\frac{\partial p}{\partial T}\right)_v$ in the case of solid and liquid bodies, it is advantageous to use the relation (11) here, from which :

$$c_p - c_v = - T\left(\frac{\partial p}{\partial v}\right)_T \cdot \left(\frac{\partial v}{\partial T}\right)_p^2 \quad . \quad . \quad . \quad (99)$$

follows.

Since $\left(\frac{\partial p}{\partial v}\right)_T$ is necessarily negative, we always have $c_p > c_v$ except in the limiting case, as for water at 4° C., at which the expansion coefficient is zero, when we have $c_p = c_v$. For solid and liquid bodies the difference $c_p - c_v$ is in general relatively small, or the ratio $\frac{c_p}{c_v} = \kappa$ is only slightly greater than 1. That is, in the case of solid and liquid bodies the dependence of the energy on the temperature plays a much greater part than the volume.

The case is different with gases, as we have seen from equation (55).

§ 52. We shall now make a further application of the second law to a physical system that represents the general type of a machine which functions with a definite period and generates mechanical work from heat. The essence of such a machine consists in executing a cycle after the completion of which no other change has occurred in nature except that a certain amount of mechanical work has been performed, say in lifting a weight, and that certain bodies that serve as heat reservoirs have given up or received heat. Such a process can, for example, be carried out by a gas which experiences a series of successive adiabatic and isothermal compressions and dilatations.

During an adiabatic change the gas remains thermally isolated; during an isothermal change it is in thermal connexion with a heat reservoir at the temperature in question. The process may, however, also be connected with changes of the aggregate state or with chemical transformations; all that is essential is that the process shall be cyclical and that at the end of a period no changes shall have remained except those above mentioned. For from each of the two principal laws it is possible to derive a relation between the different changes caused by the process.

According to the first law we have for the cycle, by (39):

$$A + \Sigma Q = 0 \quad . \quad . \quad . \quad . \quad . \quad (100)$$

Here A denotes the sum total of the external work performed, Q the heat that has been transferred from a heat reservoir to the system under consideration during the process, the summation being extended over all heat reservoirs.

According to the second law the sum of the entropies of all the heat reservoirs becomes increased as a result of the cycle. If we assume for simplicity that the heat capacities of the reservoirs are so great that the loss of the quantity of heat Q does not appreciably alter the temperature of the reservoir, the change of entropy of the reservoir due to the loss of Q amounts, by (82) and (49), to $-\frac{Q}{T}$, since on account of the constancy of T this expression also holds for finite values of Q. Accordingly we have, by the second law, summing over all the reservoirs:

$$\sum \frac{Q}{T} \leqq 0 \quad . \quad . \quad . \quad . \quad . \quad . \quad (101)$$

This was the first exact formulation of the second law and was due to Clausius.

We shall now consider the case of only two heat reservoirs, at the temperatures T_1 and $T_2 (> T_1)$ and shall assume that the process is a reversible cycle (CARNOT'S CYCLE). The two equations then run :

$$A + Q_1 + Q_2 = 0 \quad . \quad . \quad . \quad . \quad (102)$$

$$\frac{Q_1}{T_1} + \frac{Q_2}{T_2} = 0 \quad . \quad . \quad . \quad . \quad . \quad (103)$$

If the cycle is carried out in the direction which leads to the production of external work (raising a weight), then A, the external work done, is negative; so we set $A' = -A > 0$. Further, we see from (103) that Q_1 and Q_2 have opposite signs and that the absolute value of Q_2 is greater than that of Q_1. Now since A is negative, Q_2 is positive by (102) and Q_1 negative. So we set $Q'_1 = -Q_1 > 0$ and we can picture the result of the cycle in the following simple way. The heat reservoir 2 has given up the quantity of heat Q_2. Of this quantity the part Q'_1 has passed into the cooler reservoir 1; the other part $Q_2 - Q'_1 = A'$ has been transformed into mechanical work. Between these three positive quantities the following relations hold :

$$Q_2 : Q'_1 : A' = T_2 : T_1 : (T_2 - T_1)$$

which are entirely independent of the nature of the substance used in the process. In other words, by allowing a hotter reservoir at the temperature T_2 to give up a quantity of heat Q'_1 to a cooler reservoir at the temperature T_1 we can arrange a reversible cycle such that it enables us to obtain work :

$$A' = \frac{T_2 - T_1}{T_1} Q'_1 \quad . \quad . \quad . \quad . \quad (104)$$

from the hotter reservoir.

If the cyclic process is not reversible the energy equation (102) remains in force but in place of the entropy equation (103) we have the inequality :

$$\frac{Q_1}{T_1} + \frac{Q_2}{T_2} < 0$$

which, combined with (102), gives:

$$A' = -A = Q_1 + Q_2 < Q_1 - \frac{T_2}{T_1} Q_1$$

or:

$$A' < \frac{T_2 - T_1}{T_1} \cdot Q'_1$$

Comparison with (104) shows that the work to be obtained by the transition of a quantity of heat Q'_1 at the temperature T_2 to the temperature T'_1 is always less for an irreversible process than for a reversible process. Thus the latter amount, represented by (104), is the maximum amount of heat which can be obtained by means of a cyclic process with any physical system through the passage of heat Q'_1 from the hotter reservoir at the temperature T_2 to the temperature T_1.

A very special case of such a cycle is that in which heat passes directly by conduction from the hotter reservoir to the cooler reservoir. Then nothing except the two reservoirs has altered at all, and so $A = 0$. Consequently, by (100), $Q_1 + Q_2 = 0$ and by (101):

$$Q_1 \cdot \left(\frac{1}{T_1} - \frac{1}{T_2}\right) \leqq 0$$

Hence if T_1 differs from T_2, and if Q_1 does not vanish, the sign of Q_1, the heat given up by the reservoir 1, is the same as that of $T_1 - T_2$, that is, the heat passes in the direction of the higher to the lower temperature, and the process of heat conduction is irreversible, just like friction and the process of expansion when no external work is done.

Lastly, let us apply our results to an *isothermal* cycle, otherwise arbitrary, reversible or irreversible, performed on any physical system whatsoever. Then we need consider only a single heat reservoir at the constant temperature T. The equations (100) and (101) become:

$$A + Q = 0$$

and:

$$Q \leqq 0.$$

From this it follows that $A \geqq 0$, that is, work is used up and the equivalent heat is produced in the reservoir. This inequality is the analytical expression of the impossibility of a perpetual motion machine of the second kind.

If the process is reversible the sign of inequality vanishes and both the heat Q and the work A become zero. It is due to this theorem that the second law is so fruitful when applied to reversible isothermal cycles.

§ 53. We shall now leave cyclic processes in order to deal with the general question of the direction in which any change in an arbitrarily given physical-chemical configuration occurs in nature. Let us imagine any system of bodies at the same temperature T and at the same pressure p. Let us enquire into the conditions under which a physical or a chemical change occurs in these bodies. The difference between this and our earlier discussions consists in our not necessarily assuming that the system is isolated from its surroundings. Accordingly we may not assert that its entropy necessarily increases.

For an infinitely small change of state we have by the first law :

$$dU = Q - pdV \quad . \quad . \quad . \quad . \quad (104a)$$

where U denotes the total energy, V the total volume of the system and Q the heat transferred to it from outside.

According to the second law the change in the sum of the entropies of all the bodies changed by the process is :

$$dS + dS_a \geqq 0$$

where S denotes the entropy of the system, S_a the entropy of the external bodies (atmospheric air, walls of the vessel, the liquid in the calorimeter), which we assume also to be at the temperature T. Now by (82) and (49) :

$$dS_a = -\frac{Q}{T}$$

so by substituting the value of Q from (104a) :

$$dS - \frac{dU + pdV}{T} \geqq 0 \quad . \quad . \quad . \quad . \quad (105)$$

In this relation only such quantities occur as refer to the system in question itself; the influences of the external surroundings are completely eliminated. It represents the most general statement of the second law with regard to the occurrence of any physical or chemical change in a configuration.

It is to be observed that the inequality (105) by no means contradicts the equation (82). For the latter refers only to a physical change of state of a homogeneous body but the former to any physical or chemical change of any configuration whatsoever. Hence in general the expression (105) is an incomplete differential and cannot as a rule be integrated; that is, the second law does not allow us to make a general statement about a finite physical or chemical change of state of a system in the case where we do not know the external conditions to which the system is subject. This is really evident from the outset and holds equally well for the first law.

To arrive at a law for a finite change of state of a system we must know such external conditions as allow (105) to be integrated. Since the external conditions can be chosen at will, there are of course several of them, of which three, however, are distinguished by their special importance: firstly, we may completely isolate the system from its surroundings, keeping the volume V constant; secondly, we may keep the temperature T and the volume V constant (isothermal-isochoric process); thirdly, we may keep the temperature T and the pressure p constant (isothermal-isobaric process). We shall discuss these three cases in turn, of which each offers special points of interest.

§ 54. Isolated System of Constant Volume. Since $Q = 0$ as well as $V = \text{const.}$, we have from the first law also that $U = \text{const.}$, and the relation (105) gives:

$$dS \geqq 0$$

That is, the entropy of the system increases. This is the formulation of the second law which has already been obtained in § 47. But we shall here make a further deduction from it. For it furnishes us directly with a sufficient condition for the stable equilibrium of the system. If, namely, the system is in the state for which its entropy has its maximum value, no further change is possible. The absolute maximum of the entropy is therefore a sufficient condition for equilibrium. This condition is not exactly necessary; for it is possible for a system to remain unchanged, although the second law would allow a change. Since the maximum of the entropy is of course fully determined by the values of U and V, we may say that the entropy S of the configuration in the case of absolute equilibrium is a definite function of U and V. The way in which it depends on U and V is shown by equation (82) from which we get:

$$\left(\frac{\partial S}{\partial U}\right)_V = \frac{1}{T} \quad . \quad . \quad . \quad . \quad . \quad (106)$$

$$\left(\frac{\partial S}{\partial V}\right)_U = \frac{p}{T} \quad . \quad . \quad . \quad . \quad . \quad (107)$$

If, in particular, we assume a single homogeneous body, we see that its entire thermodynamic behaviour is determined by expressing S as a function of U and V. For the elimination of U from the last two equations gives p as a function of T and V, and the equation (106) alone gives U as a function of T and V.

In Part Four of the present volume (§ 125) we shall become acquainted with a method of expressing S in terms of U and V; this makes it possible to solve the principal problem of thermodynamics.

Let us calculate S for the particular case of an ideal gas. From (82), (97) and (96) we get:

$$dS = \frac{C_v dT}{T} + \frac{RndV}{V} \quad . \quad . \quad . \quad . \quad (108)$$

where C_v denotes the heat capacity of the gas at constant volume. Thus:

$$S = C_v \log T + Rn \log V + \text{const.} \quad . \quad . \quad (109)$$

and by (97):

$$S = C_v \log U + Rn \log V + \text{const.} \quad . \quad . \quad (110)$$

The integration constant depends on the chemical composition of the gas; it must remain undetermined here because the differential dS in (82) refers only to changes in U and V but not to such as involve the chemical composition.

§ 55. TEMPERATURE AND VOLUME GIVEN. Since T and V are constant the relation (105) may be written in the form:

$$d\Psi \geqq 0 \quad . \quad . \quad . \quad . \quad . \quad . \quad (111)$$

where:

$$\Psi = S - \frac{U}{T} \quad . \quad . \quad . \quad . \quad . \quad . \quad (112)$$

So in this case, too, every change of state occurs in the sense of the growth of a definite quantity; this quantity is not now the entropy, however, but the function Ψ which is characteristic for the variables T and V. Further, we can deduce from this function results similar to those deduced in the previous section from the function S. What makes this point of view so important is the fact that the temperature T can be measured in practice much more directly than the energy U and is therefore more appropriate as the independent variable.

By (111) the sufficient condition for stable equilibrium is that the function Ψ should be a maximum. Hence in stable equilibrium the quantity Ψ is a perfectly definite function of T and V. The way in which it depends on T and V results by (112) from the complete differential:

$$d\Psi = dS - \frac{dU}{T} + \frac{U}{T^2} dT$$

or, by (82):

$$d\Psi = \frac{U}{T^2} dT + \frac{p dV}{T} \quad . \quad . \quad . \quad . \quad (113)$$

Consequently:

$$\left(\frac{\partial \Psi}{\partial T}\right)_V = \frac{U}{T^2} \quad . \quad . \quad . \quad . \quad . \quad (114)$$

$$\left(\frac{\partial \Psi}{\partial V}\right)_T = \frac{p}{T} \quad . \quad . \quad . \quad . \quad . \quad (115)$$

Hence if we know the way in which the characteristic function Ψ depends on T and V, we obtain unique values of the energy and the pressure. Concerning the theoretical determination of Ψ see § 125. If we differentiate (114) with respect to V, (115) with respect to T and equate the expressions obtained, we arrive at the relation (88) already known to us. For an ideal gas we have, by integrating (113), (97) and (96), or more directly from (112) and (109):

$$\Psi = C_v \log T + Rn \log V + \frac{\text{const.}}{T} + \text{const.} \quad . \quad . \quad (116)$$

Thus the expression for Ψ contains two undetermined constants.

§ 56. Temperature and Pressure Given. This case is important because it is even easier to measure the pressure than the temperature. Here the relation (105) assumes the form:

$$d\Phi \geqq 0 \quad . \quad . \quad . \quad . \quad . \quad . \quad (117)$$

where:

$$\Phi = S - \frac{U + pV}{T} = S - \frac{W}{T} \quad . \quad . \quad (118)$$

In the latter expression we have introduced the enthalpy W (heat function at constant pressure, see § 34), by (64). Here Φ is the characteristic function and the maximum value of Φ determines the state of stable equilibrium. For the dependence of the function Φ on T and p in the state of equilibrium we get from (118):

$$d\Phi = dS - \frac{dU + pdV + Vdp}{T} + \frac{W}{T^2}dT$$

and by (82):

$$d\Phi = \frac{W}{T^2}dT - \frac{V}{T}dp \quad . \quad . \quad . \quad . \quad (119)$$

Hence:

$$\left(\frac{\partial \Phi}{\partial T}\right)_p = \frac{W}{T^2} \quad . \quad . \quad . \quad . \quad . \quad (120)$$

$$\left(\frac{\partial \Phi}{\partial p}\right)_T = -\frac{V}{T} \quad . \quad . \quad . \quad . \quad . \quad (121)$$

If we differentiate (120) with respect to p, (121) with respect to T and equate the two results, we get:

$$\left(\frac{\partial W}{\partial p}\right)_T = V - T\left(\frac{\partial V}{\partial T}\right)_p \quad . \quad . \quad . \quad (122)$$

For an ideal gas we have, by (118), (109), (96), (97) and (55):

$$\Phi = C_p \log T - Rn \log p + \frac{\text{const.}}{T} + \text{const.} \quad . \quad (123)$$

So in this case, also, two constants remain undetermined.

§ 56*a*. An example of the advantages that accrue from introducing the independent variables T and p is given by the theory of Joule and Thomson's experiment which was described in § 24; this involved the adiabatic expansion of a gas without external work being done. The theory is contained in equation (43), which, by introducing the enthalpy w of unit mass, we may write in the form $w' - w = 0$. If we now assume the difference of pressure $p' - p$ on both sides of the valve to be very small and equal to Δp, the difference in temperature $T' - T$ on both sides will also be very small ($= \Delta T$), and we then have:

$$\left(\frac{\partial w}{\partial T}\right)_p \Delta T + \left(\frac{\partial w}{\partial p}\right)_T \Delta p = 0$$

Consequently, by using (67) and (122), we get:

$$\Delta T = \frac{T\left(\frac{\partial v}{\partial T}\right)_p - v}{c_p} \cdot \Delta p \quad . \quad . \quad . \quad (124)$$

For an ideal gas the numerator on the right-hand side is equal to zero, and the temperature difference ΔT vanishes, as it should do. From this we see that the Joule-Thomson effect affords a very direct and delicate means of detecting deviations in the behaviour of a gas from that of an ideal gas. Actually, ΔT at ordinary temperatures and pressures is appreciably negative in the case of air. It is on this fact that the idea of Linde's method of liquefying air is founded. In the case of hydrogen ΔT is appreciably positive.

If we take as our equation of state that of van der Waals (19) we get from (124), for small values of a and b, to a first approximation :

$$\Delta T = \left(\frac{2a}{RT} - b\right) \cdot \frac{\Delta p}{c_p} . \quad . \quad . \quad . \quad (125)$$

a relation which agrees approximately with the results of measurements. For most gases the expression in brackets is as in the case of air positive, which corresponds to a cooling effect, since Δp is always negative. Hydrogen is an exception because a is particularly small in its case. But by means of an appropriate preliminary cooling it is even possible in the case of hydrogen to make the first term in the bracket exceed b, the second term.

§ 57. The general relationships developed above may also be formulated in other ways. One expression which is particularly distinguished by its clarity is worth mentioning. It depends on the introduction of the function :

$$F = U - TS = - T . \Psi \quad . \quad . \quad . \quad (126)$$

which, like the Ψ, may by § 55 serve as the characteristic function of the independent variables T and V. Introducing F instead of Ψ we see that the relations (114) and (115) become :

$$U = F - T\left(\frac{\partial F}{\partial T}\right)_V \quad . \quad . \quad . \quad . \quad (127)$$

$$p = - \left(\frac{\partial F}{\partial V}\right)_T \quad . \quad . \quad . \quad . \quad . \quad . \quad (128)$$

By comparing (127) and (126) we see that :

$$S = - \left(\frac{\partial F}{\partial T}\right)_V . \quad . \quad . \quad . \quad . \quad . \quad (128a)$$

The quantity F is endowed with a clear meaning owing to the fact that, as we see from (126), it represents an energy. If we now consider any *isothermal* process, the general relation (105) may be written in the following form:

$$dF \leqq - pdV \quad . \quad . \quad . \quad . \quad . \quad (129)$$

or, if we integrate from any initial state to any final state denoted by a dash:

$$F' - F \leqq A \quad . \quad . \quad . \quad . \quad . \quad (130)$$

where A denotes the mechanical work performed by external forces during the process. If the isothermal process is *reversible*, the equation is:

$$F' - F = A \quad . \quad . \quad . \quad . \quad . \quad (131)$$

and a comparison with the equation (40) of the first law shows that the function F is related to the external work A in exactly the same way as the total energy U to the sum $A + Q$ of the external work and the heat received from without. This can also be expressed as follows: in a reversible process the law (40) concerning the conservation of energy resolves into two separate laws, namely the equation (131) and the supplementary equation:

$$G' - G = Q \quad . \quad . \quad . \quad . \quad (132)$$

where:

$$G = U - F = TS \quad . \quad . \quad . \quad . \quad (132a)$$

The theorem which asserts that the mechanical equivalent of the external work is independent of the path taken from the initial to the final state (§ 18) thus does not hold only for the sum $A + Q$, but also for the individual summands A and Q.

Hence, following Helmholtz, we call F the "free" energy, G the "bound" energy. The free energy F has the same meaning for the external mechanical work as the total energy U for the sum of the work and the heat. In particular we see that for a cyclic process we have not only $A + Q = 0$ but also $A = 0$ and $Q = 0$, as has already been deduced at the end of § 52.

If the isothermal process is *irreversible* the inequality :

$$F' - F < A \quad . \quad . \quad . \quad . \quad . \quad (133)$$

holds. That is, the free energy increases less than would correspond to the external work performed, so that in this sense external work is lost. If the process occurs without external work being done, for example, at constant volume (§ 55), then $A = 0$, and :

$$F' - F < 0 \quad . \quad . \quad . \quad . \quad . \quad (134)$$

That is, the free energy decreases. The amount of this decrease may be regarded as a measure of the work of the forces (chemical relationship, affinity) which bring about the process; this work is lost as far as mechanical work is concerned. To find its amount we must carry out the same change in some reversible way. Then, by (131), the amount desired, $F' - F$, is actually obtained as external work A.

A particular example will make this clearer. If an aqueous solution of a non-volatile salt is diluted in some isothermal way by adding pure water, the heat of dilution can be negative or positive, according to the sign of $U' - U$, where U denotes the sum of the energies of the original solution and of the water that is to be added (initial state) and U' denotes the energy of the final solution. The external work arising from the simultaneous change of volume can always be neglected. On the other hand the difference $F' - F$ is always negative. To measure it we perform the isothermal process of dilution in some reversible way, such as the following. We first allow the water which is to be added to vaporize infinitely slowly. We then allow the vapour to expand further until the density of the vapour is equal to the density which saturated water vapour has in contact with the solution. We then bring the vapour into contact with the solution; the equilibrium is not disturbed by this step. Finally by compressing the vapour over the solution infinitely slowly we completely condense it. The whole process is reversible, and so by (131) the total external work done is equal to $F' - F < 0$; that is, work is gained (a weight is raised). This work is lost if the liquid water is added directly to the solution.

It is to be noted that all these theorems hold only for isothermal processes. The introduction of the free energy does not suffice for formulating more general laws.

In the expression (126) for the free energy the first term U as a rule easily predominates over the second term TS in chemical processes. For this reason we may often, especially in the case of low temperatures, regard the decrease of U, that is, the heat of transformation (*Wärmetönung*), instead of the decrease of F, as a measure of the chemical work; so we may enunciate the additional theorem that chemical changes always occur in the sense of decreasing U, that is, are accompanied by the generation of external heat (BERTHELOT'S PRINCIPLE). But at high temperatures where T and, in the case of gases and dilute solutions, S assume great values, the term TS can no longer be neglected without causing an appreciable error. Hence in such cases chemical changes often occur in the direction of increasing total energy, that is, heat is taken up from the surroundings.

14

Reprinted by permission of the publisher from pp. 31–45, 143–146, and 151 of *Natural Philosophy of Cause and Chance,* Oxford University Press, 1949, 223 pp.

ANTECEDENCE: THERMODYNAMICS

M. Born

We have now to discuss the experiences which make it possible to distinguish in an objective empirical way between past and future or, in our terminology, to establish the principle of antecedence in the chain of cause and effect. These experiences are connected with the production and transfer of heat. There would be a long story to tell about the preliminary steps necessary to translate the subjective phenomena of hot and cold into the objective language of physics: the distinction between the quality 'temperature' and the quantity 'heat', together with the invention of the corresponding instruments, the thermometer and calorimeter. I take the technical side of this development to be well known and I shall use the thermal concepts in the usual way, although I shall have to analyse them presently from the standpoint of scientific methodology. It was only natural that the measurable quantity heat was first regarded as a kind of invisible substance called caloric. The flow of heat was treated with the methods developed for material liquids, yet with one important difference: the inertia of the caloric fluid seemed to be negligible; its flow was determined by a differential equation which is not of the second but of the first order in time. It is obtained from the continuity equation (see (4.5))

$$\dot{Q}+\operatorname{div}\mathbf{q} = 0 \tag{5.1}$$

by assuming that the change of the density of heat Q is proportional to the change of temperature T, $\delta Q = c\,\delta T$ (where c is the specific heat), while the current of heat $\mathbf{q}$ is proportional to the negative gradient of temperature, $\mathbf{q} = -\kappa \operatorname{grad} T$ (where κ is the coefficient of conductivity). Hence

$$c\frac{\partial T}{\partial t} = \kappa\Delta T, \tag{5.2}$$

a differential equation of the first order in time. This equation was the starting-point of one of the greatest discoveries in mathematics, Fourier's theory of expansion of arbitrary functions in terms of orthogonal sets of simple periodic functions, the prototype of numerous similar expansions and the embryo from which a considerable part of modern analysis and mathematical physics developed.

But that is not the aspect from which we have here to regard the equation (5.2); it is this:

The equation does not allow a change of t into $-t$, the result cannot be compensated by a change of sign of other variables as happens in Maxwell's equations. Hence the solutions exhibit an essential difference of past and future, a definite 'flow of time' as one is used to say—meaning, of course, a flow of events in time. For instance, an elementary solution of (5.2) for the temperature distribution in a thin wire along the x-direction is

$$T-T_0 = \frac{C}{\sqrt{t}}e^{-(cx^2/4\kappa t)}, \tag{5.3}$$

which describes the spreading and levelling out of an initially high temperature concentrated near the point $x = 0$, an obviously irreversible phenomenon.

I do not know enough of the history of physics to understand how this theory of heat conduction was reconciled with the general conviction that the ultimate laws of physics were of the Newtonian reversible type.

Before a solution of this problem could be attempted another important step was necessary: the discovery of the equivalence of heat and mechanical work, or, as we say to-day, of the first law of thermodynamics. It is important to remember that this discovery was made considerably later than the invention of the steam-engine. Not only the production of heat by mechanical work (e.g. through friction), but also the production of work from heat (steam-engine) was known. The new feature was the statement that a given amount of heat always corresponds to a definite amount of mechanical work, its 'mechanical equivalent'. Robert Mayer pronounced this law on very scanty and indirect evidence, but obtained a fairly good value for the equivalent from known properties of gases, namely from the difference of heat necessary to raise the temperature by one degree if either the volume is kept constant or the gas allowed to do work against a constant pressure. Joule investigated the same problem by systematic experiments which proved the essential point, namely that the work necessary to transfer a system from one equilibrium state to another depends only on these two states, not on the process of application of the work. This is the real content of the first law; the determination of the numerical value of the mechanical equivalent, so much stressed in text-books, is a matter of physical technique. To get our notions clear, we have now to return to the logical and philosophical foundations of the theory of heat.

The problem is to transform the subjective sense impressions of hot and cold into objective measurable statements. The latter are, of course, again somewhere connected with sense impressions. You cannot read an instrument without looking at it. But there is a difference between this looking at, say, a thermometer with which a nurse measures the temperature of a patient and the feeling of being hot under which the patient suffers.

It is a general principle of science to rid itself as much as possible from sense qualities. This is often misunderstood as meaning elimination of sense impressions, which, of course, is absurd. Science is based on observation, hence on the use of the senses. The problem is to eliminate the subjective features and to maintain only statements which can be confirmed by several individuals in an objective way. It is impossible to explain to anybody what I mean by saying 'This thing is red' or 'This thing is hot'. The most I can do is to find out whether other persons call the same things red or hot. Science aims at a closer relation between word and fact. Its method consists in finding correlations of one kind of subjective sense impressions with other kinds, using the one as indicators for the other, and in this way establishes what is called a fact of observation.

Here I have ventured again into metaphysics. At least, a philosopher would claim that a thorough study of these methodological principles is beyond physics. I think it is again a rule of our craft as scientists, like the principle of inductive inference, and I shall not analyse it further at this moment.

In the case of thermal phenomena, the problem is to define the quantities involved—temperature, heat—by means of observable objective changes in material bodies. It turns out that the *concepts of mechanics*, configuration and force, strain and stress, suffice for this purpose, but that the *laws of mechanics* have to be essentially changed.

Let us consider for simplicity only systems of fluids, that is of continuous media, whose state in equilibrium is defined by *one* single strain quantity, the density, instead of which we can also, for a given mass, take the total volume V. There is also only *one* stress quantity, the pressure p. From the standpoint of mechanics the pressure in equilibrium is a given function of the volume, $p = f(V)$.

Now all those experiences which are connected with the subjective impression of making the fluid hotter or colder, show that this law of mechanics is wrong: the pressure can be changed at constant volume—namely 'by heating' or 'by cooling'.

Hence the pressure p can be regarded as an independent variable besides the volume V, and this is exactly what thermodynamics does.

The generalization for more complicated substances (such as those with rigidity or magnetic polarizability) is so obvious that I shall stick to the examples of fluids, characterized by two thermodynamically independent variables V, p. But it is necessary to consider systems consisting of several fluids, and therefore one has to say a word about different kinds of contact between them.

To shorten the expression, one introduces the idea of 'walls' separating different fluids. These walls are supposed to be so thin that they play no other part in the physical behaviour of the system than to define the interaction between two neighbouring fluids. We shall assume every wall to be impenetrable to matter, although in theoretical chemistry semi-permeable partitions are used with great advantage. Two kinds of walls are to be considered.

An *adiabatic wall* is defined by the property that equilibrium of a body enclosed by it is not disturbed by any external process as long as no part of the wall is moved (distance forces being excluded in the whole consideration).

Two comments have to be made. The first is that the adiabatic property is here defined without using the notion of heat; that is essential, for as it is our aim to define the thermal concepts in mechanical terms, we cannot use them in the elementary definitions. The second remark is that adiabatic enclosure of a system can be practically realized, as in the Dewar vessel or thermos flask, with a high degree of approximation. Without this fact, thermodynamics would be utterly impracticable.

The ordinary presentations of this subject, though rather careless in their definitions, cannot avoid the assumption of the possibility of isolating a system thermally; without this no calorimeter would work and heat could not be measured.

The second type of wall is the *diathermanous wall*, defined by the following property: if two bodies are separated by a diathermanous wall, they are not in equilibrium for arbitrary values of their variables p_1, V_1 and p_2, V_2, but only if a definite relation between these four quantities is satisfied

$$F(p_1, V_1, p_2, V_2) = 0. \tag{5.4}$$

This is the expression of thermal contact; the wall is only introduced to symbolize the impossibility of exchange of material.

The concept of temperature is based on the experience that two bodies, being in thermal equilibrium with a third one, are also in thermal equilibrium with another. If we write (5.4) in the short form $F(1, 2) = 0$, this property of equilibrium can be expressed by saying that of the three equations

$$F(2, 3) = 0, \qquad F(3, 1) = 0, \qquad F(1, 2) = 0, \tag{5.5}$$

any two always involve the third. This is only possible if (5.4) can be brought into the form

$$f_1(p_1, V_1) = f_2(p_2, V_2). \tag{5.6}$$

Now one can use one of the two bodies, say 2, as thermometer and introduce the value of the function

$$f_2(p_2, V_2) = \vartheta \tag{5.7}$$

as *empirical temperature*. Then one has for the other body the so-called *equation of state*

$$f_1(p_1, V_1) = \vartheta. \tag{5.8}$$

Any arbitrary function of ϑ can be chosen as empirical temperature with equal right; the choice is restricted only by practical considerations. (It would be impractical to use a thermometric substance for which two distinguishable states are in thermal equilibrium.) The curves $\vartheta =$ const. in the pV-plane are independent of the temperature scale; they are called *isotherms*.

It is not superfluous to stress the extreme arbitrariness of the temperature scale. Any suitable property of any substance can be chosen as thermometric indicator, and if this is done, still the scale remains at our disposal. If we, for example, choose a gas at low pressures, because of the simplicity of the isothermal compression law $pV =$ const., there is no reason to take $pV = \vartheta$ as measure of temperature: one could just as well take $(pV)^2$ or $\sqrt{(pV)}$. The definition of an 'absolute' scale of temperature was therefore an urgent problem which was solved by the discovery of the second law of thermodynamics.

The second fundamental concept of thermodynamics, that of heat, can be defined in terms of mechanical quantities by a proper interpretation of Joule's experiments. As I have pointed out already, the gist of these experiments lies in the following fact: If a body in an adiabatic enclosure is brought from one (equilibrium) state to another by applying external work, the amount of this work is always the same in whatever form (mechanical, electrical, etc.) and manner (slow or fast, etc.) it is applied.

Hence for a given initial state (p_0, V_0) the work done adiabatically is a function U of the final state (p, V), and one can write

$$W = U - U_0; \tag{5.9}$$

the function $U(p, V)$ is called the *energy* of the system. It is a quantity directly measurable by mechanical methods.

If we now consider a non-adiabatic process leading from the initial state (p_0, V_0) to the final state (p, V), the difference $U - U_0 - W$ will not be zero, but can be determined if the energy function $U(p, V)$ is known from previous experiment. This difference

$$U-U_0-W = Q \tag{5.10}$$

is called the *heat* supplied to the system during the process. Equation (5.10) is the definition of heat in terms of mechanical quantities.

This procedure presupposes that mechanical work is measurable however it is applied; that means, for example, that the displacements of and the forces on the surface of a stirring-wheel in a fluid, or the current and resistance of a wire heating the fluid, must be registered even for the most violent reactions. Practically this is difficult, and one uses either stationary processes of a comparatively long duration where the irregular initial and final stages can be neglected (this includes heating by a stationary current), or extremely slow, 'quasi-static' processes; these are in general (practically) reversible, since no kinetic energy is produced which could be irreversibly destroyed by friction. In ordinary thermodynamics one regards every curve in the pV-plane as the diagram of a reversible process; that means that one allows infinitely slow heating or cooling by bringing the system into thermal contact with a series of large heat reservoirs which differ by small amounts of temperature. Such an assumption is artificial; it does not even remotely correspond to a real experiment. It is also quite superfluous. We can restrict ourselves to adiabatic quasi-static processes, consisting of slow movements of the (adiabatic) walls. For these the work done on a simple fluid is

$$dW = -p\,dV, \tag{5.11}$$

where p is the equilibrium pressure, and the first theorem of thermodynamics (5.10) assumes the form

$$dQ = dU+p\,dV = 0. \tag{5.12}$$

For systems of fluids separated by adiabatic or diathermanous walls the energy and the work done are additive (according to our definition of the walls); hence, for instance,

$$dQ = dQ_1+dQ_2 = dU+p_1\,dV_1+p_2\,dV_2, \tag{5.13}$$

where

$$U = U_1+U_2.$$

This equation is of course only of interest for the case of thermal contact where the equation (5.6) holds; the system has then only three independent variables, for which one can choose V_1, V_2 and the temperature ϑ, defined by (5.7) and (5.8). Then $U_1 = U_1(V_1, \vartheta)$, $U_2 = U_2(V_2, \vartheta)$, and (5.13) takes the form

$$dQ = \left(\frac{\partial U_1}{\partial V_1}+p_1\right)dV_1+\left(\frac{\partial U_2}{\partial V_2}+p_2\right)dV_2+\left(\frac{\partial U_1}{\partial \vartheta}+\frac{\partial U_2}{\partial \vartheta}\right)d\vartheta = 0. \tag{5.14}$$

Every adiabatic quasi-static process can be represented as a line in the three-dimensional $V_1 V_2 \vartheta$-space which satisfies this equation; let us call these for brevity 'adiabatic lines'.

Equation (5.14) is a differential equation of a type studied by Pfaff. Pfaffian equations are the mathematical expression of elementary thermal experiences, and one would expect that the laws of thermodynamics are connected with their properties. That is indeed the case, as Carathéodory has shown. But classical thermodynamics proceeded in quite a different way, introducing the conception of idealized thermal machines which transform heat into work and vice versa (William Thomson—Lord Kelvin), or which pump heat from one reservoir into another (Clausius). The second law of thermodynamics is then derived from the assumption that not all processes of this kind are possible: you cannot transform heat completely into work, nor bring it from a state of lower temperature to one of higher 'without compensation' (see Appendix, **6**). These are new and strange conceptions, obviously borrowed from engineering. I have mentioned that the steam-engine existed before thermodynamics; it was a matter of course at that time to use the notions and experiences of the engineer to obtain the laws of heat transformation, and the establishment of the abstract concepts of entropy and absolute temperature by this method is a wonderful achievement. It would be ridiculous to feel anything but admiration for the men who invented these methods. But even as a student, I thought that they deviated too much from the ordinary methods of physics; I discussed the problem with my mathematical friend, Carathéodory, with the result that he analysed it and produced a much more satisfactory solution. This was about forty years ago, but still all textbooks reproduce the 'classical' method, and I am almost certain that the same holds for the great majority of lectures—I know, however, a few exceptions, namely those of the late R. H. Fowler and his school. This state of affairs seems to me one of unhealthy conservatism. I take in these lectures an opportunity to advocate a change.

The central point of Carathéodory's method is this. The principles from which Kelvin and Clausius derived the second law are formulated in such a way as to cover the greatest possible range of processes incapable of execution: in no way whatever can heat be completely transformed into work or raised to a higher level of temperature. Carathéodory remarked that it is perfectly sufficient to know the existence of *some* impossible processes to derive the second law. I need hardly say that this is a logical advantage. Moreover, the impossible processes are already obtained by scrutinizing Joule's experiments a little more carefully. They consisted in bringing a system in an adiabatic enclosure from one equilibrium state to another by doing external work: it is an elementary experience, almost obvious, that you cannot get your work back by reversing the process. And that holds however near the two states are. One can therefore say that there exist adiabatically inaccessible states in any vicinity of a given state. That is Carathéodory's principle.

In particular, there are neighbour states of any given one which are inaccessible by quasi-static adiabatic processes. These are represented by adiabatic lines satisfying the Pfaffian equation (5.14). Therefore the question arises: Does Carathéodory's postulate hold for any Pfaffian or does it mean a restriction?

The latter is the case, and it can be seen by very simple mathematics indeed, of which I shall give here a short sketch (see Appendix, **7**).

Let us first consider a Pfaffian equation of two variables, x and y,

$$dQ = X\,dx + Y\,dy, \tag{5.15}$$

where X, Y are functions of x, y. This is equivalent to the ordinary differential equation

$$\frac{dy}{dx} = -\frac{X}{Y}, \tag{5.16}$$

which has an infinite number of solutions $\phi(x,y) = \text{const.}$, representing a one-parameter set of curves in the (x,y)-plane. Along any of these curves one has

$$d\phi = \frac{\partial\phi}{\partial x}dx + \frac{\partial\phi}{\partial y}dy = 0, \tag{5.17}$$

and this must be the same condition as the given Pfaffian; hence one must have

$$dQ = \lambda\, d\phi. \tag{5.18}$$

Each Pfaffian dQ of two variables has therefore an 'integrating denominator' λ, so that dQ/λ is a total differential.

For Pfaffians of three (or more) variables,

$$dQ = X\,dx + Y\,dy + Z\,dz \tag{5.19}$$

this does not hold. It is easy to give analytical examples (see Appendix, **7**); but one can see it geometrically in this way: if in (5.19) dx, dy, dz are regarded as finite differences $\xi - x, \eta - y, \zeta - z$, it is the equation of a plane through the point x, y, z; one has a plane through each point of space, continuously varying in orientation with the position of this point. Now if a function ϕ existed, these planes would have to be tangential to the surfaces $\phi(x,y,z) = \text{const.}$ But one can construct continuously varying sets of planes which are not 'integrable', i.e. tangential to a set of surfaces. For example, take all circular screws with the same axis, but varying radius and pitch, and construct at each point of every screw the normal plane; these obviously form a non-integrable set of planes.

Hence all Pfaffians can be separated into two classes: those of the form $dQ = \lambda\, d\phi$, which have an 'integrating denominator' and represent the tangential planes of a set of surfaces $\phi = \text{const.}$, and those which lack this property.

Now in the first case, $dQ = \lambda\, d\phi$, any line satisfying the Pfaffian equation (5.19) must lie in the surface $\phi = \text{const.}$ Hence an arbitrary pair of points P_0 and P in the xyz-space cannot be connected by such a line. This is quite elementary. Not quite so obvious is the inverse statement which is used in the thermodynamic application: If there are points P in any vicinity of a given point P_0 which cannot be connected with P_0 by a line satisfying the Pfaffian equation (5.19), then there exists an integrating denominator and one has $dQ = \lambda\, d\phi$.

One can intuitively understand this theorem by a continuity consideration: All points P inaccessible from P_0 will fill a certain volume, bound by a surface of accessible points going through P_0. Further, to each inaccessible point there corresponds another one in the opposite direction; hence the boundary surface must contain all accessible points: which proves the existence of the function ϕ, so that $dQ = \lambda\, d\phi$ (see Appendix, **7**).

The application of this theorem to thermodynamics is now simple. Combining it with Carathéodory's principle, one has for any two systems

$$dQ_1 = \lambda_1\, d\phi_1, \qquad dQ_2 = \lambda_2\, d\phi_2, \tag{5.20}$$

and for the combined system

$$dQ = dQ_1 + dQ_2 = \lambda\, d\phi; \tag{5.21}$$

hence

$$\lambda\, d\phi = \lambda_1\, d\phi_1 + \lambda_2\, d\phi_2. \tag{5.22}$$

Consider in particular two simple fluids in thermal contact; then the system has three independent variables V_1, V_2, ϑ, which can be replaced by $\phi_1, \phi_2, \vartheta$. Then (5.22) shows that ϕ depends only on ϕ_1, ϕ_2, and not on ϑ, while

$$\frac{\partial\phi}{\partial\phi_1} = \frac{\lambda_1}{\lambda}, \qquad \frac{\partial\phi}{\partial\phi_2} = \frac{\lambda_2}{\lambda}. \tag{5.23}$$

Hence these quotients are also independent of ϑ,

$$\frac{\partial}{\partial\vartheta}\frac{\lambda_1}{\lambda} = 0, \qquad \frac{\partial}{\partial\vartheta}\frac{\lambda_2}{\lambda} = 0,$$

from which one infers

$$\frac{1}{\lambda_1}\frac{\partial\lambda_1}{\partial\vartheta} = \frac{1}{\lambda_2}\frac{\partial\lambda_2}{\partial\vartheta} = \frac{1}{\lambda}\frac{\partial\lambda}{\partial\vartheta}. \tag{5.24}$$

Now λ_1 is a variable of the first fluid only, therefore only dependent on ϕ_1 and ϑ; in the same way $\lambda_2 = \lambda_2(\phi_2, \vartheta)$. The first equality (5.24) can only hold if both quantities depend only on ϑ. Hence

$$\frac{\partial \log \lambda_1}{\partial \vartheta} = \frac{\partial \log \lambda_2}{\partial \vartheta} = \frac{\partial \log \lambda}{\partial \vartheta} = g(\vartheta), \tag{5.25}$$

where $g(\vartheta)$ is a universal function, namely numerically identical for different fluids and for the combined system.

This simple consideration leads with ordinary mathematics to the existence of a universal function of temperature. The rest is just a matter of normalization. From (5.25) one finds for each system

$$\log \lambda = \int g(\vartheta)\, d\vartheta + \log \Phi, \qquad \lambda = \Phi e^{\int g(\vartheta)\, d\vartheta}, \tag{5.26}$$

where Φ depends on the corresponding ϕ.

If one now defines

$$\left.\begin{aligned} T(\vartheta) &= C e^{\int g(\theta)\, d\vartheta}, \\ S(\phi) &= \frac{1}{C} \int \Phi(\phi)\, d\phi, \end{aligned}\right\} \tag{5.27}$$

where the constant C can be fixed by prescribing the value of $T_1 - T_2$ for two reproducible states of some normal substance (e.g. $T_1 - T_2 = 100^\circ$, if T_1 corresponds to the boiling-point, T_2 the freezing-point of water at 1 atmosphere of pressure), then one has

$$dQ = \lambda\, d\phi = T\, dS. \tag{5.28}$$

T is the thermodynamical or absolute temperature and S the entropy.

Equation (5.28) refers only to quasi-static processes, that is, to sequences of equilibrium states. To get a result about real dynamical phenomena one has to apply Carathéodory's principle again, considering a finite transition from an initial state V_1^0, V_2^0, S^0 to a final state V_1, V_2, S. One can reach the latter one in two steps: first changing the volume quasi-statically (and adiabatically) from V_1^0, V_2^0 to V_1, V_2, the entropy remaining constant, equal to S^0, and then changing the state adiabatically, but irreversibly (by stirring, etc.) at constant volume, so that S^0 goes over into S.

Now if any neighbouring value S of S^0 could be reached in this way, one would have a contradiction to Carathéodory's principle, as the volumes are of course arbitrarily changeable. Hence for each such process one must have either $S \geqslant S^0$ or $S \leqslant S^0$. Continuity demands that the same sign holds for all initial states; it holds also for different substances since the entropy is additive (as can be easily seen). The actual sign $\geqslant$ or $\leqslant$ depends on the choice of the constant C in (5.27); if this is chosen so that T is positive, a single experience, say with a gas, shows that entropy never decreases.

It may not be superfluous to add a remark on the behaviour of entropy for the case of conduction of heat. As thermodynamics has to do only with processes where the initial and final states are equilibria, stationary flow cannot be treated: one can only ask, What is the final state of two initially separated bodies brought into thermal contact? The difficulty is that a change of entropy is only defined by quasi-static adiabatic processes; the sudden change of thermal isolation into contact, however, is discontinuous and the processes inside the system not controllable. Yet one can reduce this process to the one considered before. By quasi-static adiabatic changes of volume the temperatures can be made equal without change of entropy; then contact can be made without discontinuity, and the initial volumes quasi-statically restored, again without a change of entropy. The situation is now the same as in the initial state considered before, and it follows that any process leading to the final state must increase the entropy.

The whole chain of considerations can be generalized for more complicated systems without any difficulty. One has only to assume that all independent variables except one are of the type represented by the volume, namely arbitrarily changeable.

If one has to deal, as in chemistry, with substances which are mixtures of different components, one can regard the concentrations of these as arbitrarily variable with the help of semipermeable walls and movable pistons (see Appendix, **8**).

By using thermodynamics a vast amount of knowledge has been accumulated not only in physics but in the borderland sciences of physico-chemistry, metallurgy, mineralogy, etc. Most of it refers to equilibria. In fact, the expression 'thermodynamics' is misleading. The only dynamical statements possible are concerned with the irreversible transitions from one equilibrium state to another, and they are of a very modest character, giving the total increase of entropy or the decrease of free energy $F = U - TS$. The irreversible process itself is outside the scope of thermodynamics.

The principle of antecedence is now satisfied; but this gain is paid for by the loss of all details of description which ordinary dynamics of continuous media supplies.

Can this not be mended? Why not apply the methods of Cauchy to thermal processes, by treating each volume element as a small thermodynamical system, and regarding not only strain, stress, and energy, but also temperature and entropy as continuous functions in space? This has of course been done, but with limited success. The reason is that thermodynamics is definitely connected with walls or enclosures. We have used the adiabatic and diathermanous variety, and mentioned semi-permeable walls necessary for chemical separations; but a volume element is not surrounded by a wall, it is in free contact with its neighbourhood. The thermodynamic change to which it is subject depends therefore on the flux of energy and material constituents through its boundary, which themselves cannot be reduced to mechanics. In some limiting cases, one has found simple solutions. For instance, when calculating the velocity of sound in a gas, one tried first for the relation between pressure p and density ρ the isothermal law $p = c\rho$ where c is a constant, but found no agreement with experiment; then one took the adiabatic law $p = c\rho^{\gamma}$ where γ is the ratio of the specific heats at constant pressure and constant volume (see Appendix, **9**), which gave a much better result. The reason is that for fast vibrations there is no time for heat to flow through the boundary of a volume element which therefore behaves as if it were adiabatically enclosed. But by making the vibrations slower and slower, one certainly gets into a region where this assumption does not hold any more. Then conduction of heat must be taken into account. The hydrodynamical equations and those of heat conduction have to be regarded as a simultaneous system. In this way a descriptive or phenomenological theory can be developed and has been developed. Yet I am unable to give an account of it, as I have never studied it; nor have the majority of physicists shown much interest in this kind of thing. One knows that any flux of matter and energy can be fitted into Cauchy's general scheme, and there is not much interest in doing it in the most general way. Besides, each effect needs separate constants—e.g. in liquids compressibility, specific heat, conductivity of heat, constants of diffusion ; in solids elastic constants and parameters describing plastic flow, etc., and very often these so-called constants turn out to be not constants, but to depend on other quantities (see Appendix, **10**).

Therefore one can rightly say that with ordinary thermodynamics the descriptive method of physics has come to its natural end. Something new had to appear.

[*Editor's Note:* Material has been omitted at this point.]

6. (V. p. 38.) **On classical and modern thermodynamics**

It is often said that the classical derivation of the second law of thermodynamics is much simpler than Carathéodory's as it needs less abstract conceptions than Pfaffian equations. But this objection is quite wrong. For what one has to show is the existence of an integrating denominator of dQ. This is trivial for a Pfaffian of two variables (representing, for example, a single fluid with V, ϑ); it must be shown not to be trivial and even, in general, wrong for Pfaffians with more than two variables (e.g. two fluids in thermal contact with V_1, V_2, ϑ). Otherwise, the student cannot possibly understand what the fuss is all about. But that means explaining to him the difference between the two classes of Pfaffians of three variables, the integrable ones and the non-integrable ones. Without that all talk about Carnot cycles is just empty verbiage. But as soon as one has this difference, why not then use the simple criterion of accessibility from neighbouring points, instead of invoking quite new ideas borrowed from engineering? I think a satisfactory lecture or text-book should bring this classical reasoning as a corollary of historical interest, as I have suggested long ago in a series of papers (*Phys. Zeitschr.* **22**, pp. 218, 249, 282 (1921)).

Since writing the text I have come across one book which gives a short account of Carathéodory's theory, H. Margenau and G. M. Murphy, *The Mathematics of Physics and Chemistry* (D. van Nostrand Co., New York, 1943), § 1.15, p. 26. But though the mathematics is correct, it does not do justice to the idea. For it says on p. 28: 'This formal mathematical consequence of the properties of the Pfaff equation [namely the theorem proved in the next section of the appendix] is known as the principle of Carathéodory. It is exactly what we need for thermodynamics.' Carathéodory's principle is, of course, not that formal mathematical theorem but the induction from observation that there are inaccessible states in any neighbourhood of a given state.

7. (V. p. 39.) **Theorem of accessibility**

An example of a Pfaffian which has no integrating denominator (by the way, the same example as described in geometrical terms in the text) is this:

$$dQ = -y\,dx + x\,dy + k\,dz,$$

where k is a constant. If it were possible to write dQ in the form $\lambda d\phi$, where λ and ϕ are functions of x, y, z, one would have

$$\frac{\partial\phi}{\partial x} = -\frac{y}{\lambda}, \qquad \frac{\partial\phi}{\partial y} = \frac{x}{\lambda}, \qquad \frac{\partial\phi}{\partial z} = \frac{k}{\lambda},$$

hence

$$\frac{\partial^2\phi}{\partial y\partial z} = \frac{\partial}{\partial z}\left(\frac{x}{\lambda}\right) = \frac{\partial}{\partial y}\left(\frac{k}{\lambda}\right), \qquad \frac{\partial^2\phi}{\partial z\partial x} = -\frac{\partial}{\partial z}\left(\frac{y}{\lambda}\right) = \frac{\partial}{\partial x}\left(\frac{k}{\lambda}\right),$$

$$\frac{\partial^2\phi}{\partial x\partial y} = -\frac{\partial}{\partial y}\left(\frac{y}{\lambda}\right) = \frac{\partial}{\partial x}\left(\frac{x}{\lambda}\right),$$

or

$$x\frac{\partial\lambda}{\partial z} = k\frac{\partial\lambda}{\partial y}, \qquad -y\frac{\partial\lambda}{\partial z} = k\frac{\partial\lambda}{\partial x}, \qquad 2\lambda = x\frac{\partial\lambda}{\partial x} + y\frac{\partial\lambda}{\partial y}.$$

By substituting $\partial\lambda/\partial x$ and $\partial\lambda/\partial y$ from the first two equations in the third one finds

$$\lambda = 0.$$

Examples like this show clearly that the existence of an integrating denominator is an exception.

We now give the proof of the theorem of accessibility. Consider the solutions of the Pfaffian

$$dQ = X\,dx + Y\,dy + Z\,dz = 0, \tag{1}$$

which lie in a given surface S,

$$x = x(u, v), \qquad y = y(u, v), \qquad z = z(u, v).$$

They satisfy a Pfaffian

$$dQ = U\,du + V\,dv = 0, \tag{2}$$

where

$$U = X\frac{\partial x}{\partial u} + Y\frac{\partial y}{\partial u} + Z\frac{\partial z}{\partial u},$$

$$V = X\frac{\partial x}{\partial v} + Y\frac{\partial y}{\partial v} + Z\frac{\partial z}{\partial v}.$$

Hence through every point P of S there passes one curve, because (2) is equivalent to the ordinary differential equation

$$\frac{du}{dv} = -\frac{V}{U},$$

which has a one-parameter set of solutions $\phi(u, v) = \text{const.}$, covering the surface S.

Let us now suppose that, in the neighbourhood of a point P, there are inaccessible points; let Q be one of these. Construct through

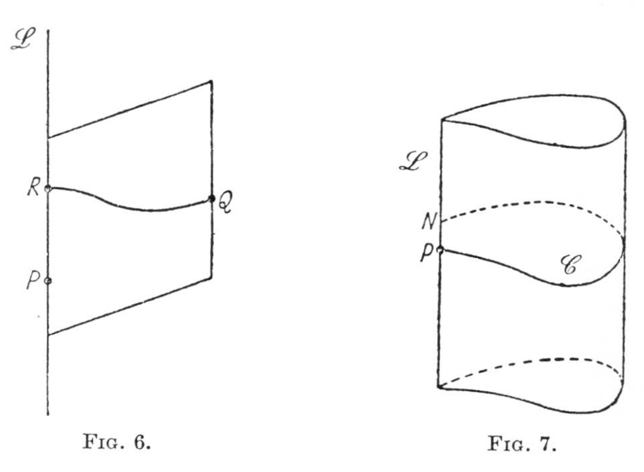

FIG. 6. FIG. 7.

P a straight line $\mathscr{L}$, which is not a solution of (1), and the plane π through Q and $\mathscr{L}$. In π there is just one curve satisfying (1) and going through Q; this curve will meet the line $\mathscr{L}$ at a point R. Then R must be inaccessible from P; for if there should exist a solution leading from P to R, then one could also reach Q from P by a continuous (though kinked) solution curve, which contradicts the assumption that Q is inaccessible from P. The point R can be made to lie as near to P as one wishes by choosing Q near enough to P.

Now we move the straight line $\mathscr{L}$ parallel to itself in a cyclic way so that it describes a closed cylinder. Then there exists on this cylinder a solution curve $\mathscr{C}$ which starts from P on $\mathscr{L}$ and meets $\mathscr{L}$ again at a point N. It follows that N and P must coincide. For otherwise one could, by deforming the cylinder, make N sweep along the line $\mathscr{L}$ towards P and beyond P. Hence there would be an interval of accessible points (like N) around P, while it has been proved before that there are inaccessible points Q in any neighbourhood of P.

As N now coincides with P the connecting curve $\mathscr{C}$ can be made, by steady deformation of the cylinder, to describe a surface which contains obviously all solutions starting from P. If this surface is given by $\phi(x, y, z) = 0$, one has

$$dQ = \lambda\, d\phi,$$

which is the theorem to be proved.

The function ϕ and the factor λ are not uniquely determined; if ϕ is replaced by $\Phi(\phi)$ one has

$$\lambda\, d\phi = \Lambda d\Phi \text{ with } \lambda = \Lambda \frac{d\Phi}{d\phi}.$$

[*Editor's Note:* Material has been omitted at this point.]

10. (V. p. 45.) **Thermodynamics of irreversible processes**

Since I wrote this section of the text a new development of the descriptive or phenomenological theory has come to my knowledge which is remarkable enough to be mentioned.

It started in 1931 with a paper by Onsager in which the attempt was made to build up a thermodynamics of irreversible processes by taking from the kinetic theory one single result, called the theorem of microscopic reversibility, and to show that this suffices to obtain some important properties of the flow of heat, matter, and electricity. The starting-point is Einstein's theory of fluctuations (see Appendix, **20**), where the relation $S = k \log P$ between probability P and entropy S is reversed, using the known dependence of S on observable quantities to determine the probability P of small deviations from equilibrium. Then it is assumed that the law for the decay of an accidental accumulation of some quantity (mass, energy, temperature, etc.) is the same as that for the flow of the same quantity under artificially produced macroscopic conditions. This, together with the reversibility theorem mentioned, determines the main features of the flow. The theory has been essentially improved by Casimir and others, amongst whom the book of Prigogine, from de Donders's school of thermodynamics in Brussels, must be mentioned. Here is a list of the literature:

L. Onsager, *Phys. Rev.* **37**, p. 405 (1931); **38**, p. 2265 (1931).

H. B. G. Casimir, Philips Research Reports, **1**, 185–96 (April 1946); *Rev. Mod. Physics*, **17**, p. 343 (1945).

C. Eckhart, *Phys. Rev.* **58**, pp. 267, 269, 919, 924 (1940).

J. Meixner, *Ann. d. Phys.* (v), **39**, p. 333 (1941); **41**, p. 409 (1943); **43**, p. 244 (1943); *Z. phys. Chem.* B, **53**, p. 235 (1943).

S. R. de Groot, *L'Effet Soret*, thesis, Amsterdam (1945); *Journal de Physique*, no. 6, p. 191.

I. Prigogine, *Étude thermodynamique des phénomènes irréversibles* (Paris, Dunod; Liége, Desoer, 1947).

[*Editor's Note:* Material has been omitted at this point.]

Part IV

SECOND THOUGHTS

Editor's Comments on Paper 15

15 MEIXNER
On the Foundation of Thermodynamics of Processes

Clausius's successors questioned the rigor of his and Lord Kelvin's derivations, the process of correction culminating with Born and Carathéodory. In turn, their successors began to question the adequacy of the thus created science of *thermostatics* for for the systematic study of irreversible processes. A realistic view of any irreversible process must recognize that the system performing it traverses a sequence of nonequilibrium states. Thus the question of how to relate the concept of *state* to nonequilibrium poses itself and is answered by the statement that continuum descriptions are needed. Furthermore, even though we believe that the second law should continue to apply, we are baffled by the technical requirement that the concept of entropy (and if we come to ponder the matter in detail, also those of energy and temperature) have so far been associated with equilibrium states only. In short, the matter of the foundations of *thermodynamics* must be faced and thought through from the start.

Meixner's paper gives a clear and convincing discussion of the principal difficulties involved, relates them to statistical thermodynamics, and opens a discussion (to be sure, not for the first time) of the interpretation that is to be given to the so-called Clausius–Duhem inequality

$$\rho \left(\frac{ds}{dt}\right) + \nabla \cdot (q/T) \geqslant 0, \tag{1}$$

in view of the fact that no unique entropy can be assigned to nonequilibrium states. The matter is not settled here conclusively, but it is hoped that these and associated considerations will attune the reader to the fact that something new is required. These new ideas will be traced in our forthcoming volume, *Irreversible Processes.*

15

Reprinted from pp. 37–47 of *A Critical Review of Thermodynamics,* E. B. Stuart, B. Gal-Or, and A. J. Brainard, eds., Mono Book Corp., Baltimore, 1970, 544 pp.

ON THE FOUNDATION OF THERMODYNAMICS OF PROCESSES

J.Meixner

Rheinisch-Westfälische Technische Hochschule, Aachen

Thermostatics* is a well-established subject with many successful applications during the hundred years of its existence. It has also found an excellent corroboration by statistical mechanics. The concepts of the thermodynamic temperature and entropy are well defined without any ambiguity and it is well known how numerical values of all sorts of variables, in particular of the entropy as a function of state, can be obtained by appropriate measurements.

The situation is very different in the thermodynamics of processes. Although there is a long history of irreversible thermodynamics, its foundation has not received sufficient attention. Since the time of Clausius' work, it has been taken for granted that, in nonequilibrium states, a unique entropy exists and there was not much discussion either about what temperature in nonequilibrium states means.

A careful study of the thermodynamics of electrical networks [1] has given considerable insight into these problems and also produced a very interesting result: the nonexistence of a unique entropy value in a state which is attained during an irreversible process.

A closely related result has been obtained by Schlögl [2] on the basis of statistical and information theoretical arguments. It states that the entropy of a system is not an objective quantity but also depends upon the observer through the information he has about the system.

We will present some critical remarks on the entropy in nonequilibrium and then find out how an entropy-free thermodynamic can be developed. The relation of this theory to classical irreversible thermodynamics and to the special constitutive assumptions of Coleman and Mizel [3] will be discussed.

However, we shall not give an analysis of the temperature concept in nonequilibrium. And we shall confine our attention to fluid materials without diffusion of material components.

* The word thermostatics will be used for what is usually called thermodynamics or equilibrium thermodynamics. Nonequilibrium thermodynamics should be called just thermodynamics. In order to avoid confusion with previous use of this word, we shall speak of thermodynamics of processes.

1. WHAT IS ENTROPY?

The entropy concept in thermostatics can be developed from the fundamental principle that heat cannot be converted into work without other changes in the system. Instead, one can proceed from the mere experimental fact that the differential $dU + P\,dV$ (or similar differentials in the case of solids or in the presence of electromagnetic fields) has an integrating factor, which depends only upon the empirical temperature, and which on proper normalization is the same for all materials. It is the reciprocal of the thermodynamic temperature. Integration then gives the entropy as a function of state. With this definition, the entropy concept is separated from the impossibility of a perpetuum mobile of second kind, or of Caratheodory's principle.

Statistical mechanics provides a direct method of evaluating the thermostatic entropy for specific systems and, in cases in which a numerical computation of the thermostatic entropy is possible, one has full agreement with experimental data.

In nonequilibrium thermodynamics (or thermodynamics of processes) no definition of an entropy has ever been given and it was only postulated that also during a process a unique entropy exists which has the well-known growth property expressed in the Clausius' inequality

$$T\,dS \geq \delta Q. \tag{1}$$

The existence of a unique nonequilibrium entropy can, however, be disproved by an analysis of the most transparent example of thermodynamics of processes, which is given by linear passive electrical networks [1].

In kinetic theories, H theorems are known, and they are interpreted as balance equations for the entropy. But kinetic theory of gases, as an example, can be developed for one- or two-, etc., particle functions, and entropies evaluated for the same process in its description with one- or two-, etc., particle functions are not the same. In particular, an N-particle function for a gas with N particles gives no entropy production, while the n-particle function with $n < N$ does.

2. WHAT IS THE SECOND LAW?

We mean here this part of the entropy law which refers to irreversible processes. Since Clausius' time it is formulated by the inequality (1). This inequality we consider as questionable because, as pointed out before, it contains a nonequilibrium entropy which is neither defined nor can be uniquely defined.

This inequality has been given by Clausius without motivation. But formerly Clausius gave another inequality

$$S(B) - S(A) > \int_A^B (\delta Q/T), \tag{2}$$

which refers to an irreversible process that starts from an equilibrium state A and ends in an equilibrium state B. The quantity $S(B) - S(A)$ is the difference of the well-defined thermostatic entropies in the equilibrium states A and B. We shall consider the inequality (2) as the expression of the Second Law for systems which do not exchange material components with their surroundings.

The inequality (1) has found another formulation

$$\rho(ds/dt) + \nabla \cdot \mathbf{S} \geq 0, \tag{3}$$

with s = specific entropy, $\mathbf{S}$ = entropy flow which, in the absence of diffusion, is the heat flow divided by the temperature. Our objection turns also against this so-called Clausius inequality for the same reason.

Now, this inequality (3) has been widely used in all field theories of irreversible processes which are the classical thermodynamics of irreversible processes, and the more recent work which was started by Coleman [4]. One cannot deny that both theories provide a good description of various phenomena. In the light of our assertion about the indeterminacy of the nonequilibrium entropy, one has to consider both theories as resting on poor foundation. On the other hand, the success of these theories forces us either to make a better foundation or to provide an understanding why these theories are not so wrong at all. This will be done by constructing a more general theory of thermodynamics of processes which contains as special cases the above-mentioned two theories.

3. PHYSICAL AND MATHEMATICAL CONTINUUM

We enunciate a few postulates before we try to develop a new thermodynamic theory of processes.

Postulate 1. A physical continuum can be treated as if it were an ideal mathematical continuum. Its theory will be called continuum mathematics.

This postulate has been tacitly used in all classical papers on continuum physics (hydrodynamics and elasticity).

Postulate 2. The physical continuum behaves approximately like a mathematical continuum if proper restrictions on the speed of the processes and on the gradients of various variables are made.

Such restrictions imply that the atomistic structure is practically smeared out.

Postulate 3. In continuum mathematics there exist fields of density, specific internal energy, velocity, stress–tensor, etc., which are smooth on the scales that are given by the atomistic structure of the corresponding physical system.

Postulate 4. There exists heat baths which can be applied over the surface of a body with the property that their relaxation times are infinitesimally small—i.e., the heat baths are constantly in thermostatic equilibrium, although their temperatures may vary in course of time.

Of course, such heat baths do not really exist. But it may suffice that the relaxation times are small on the time scale of the occurring irreversible processes.

Postulate 5. In continuum mathematics, inside the body there exists a temperature field $T(\mathbf{x}, t)$ which on the surface of the body assumes the temperatures of the respective heat baths. The values of the field $T(\mathbf{x}, t)$ on the surface of a part of the body serve as temperatures of fictive heat baths applied to the surface of this part of the body.

We note that no operational definition of the temperature $T(\mathbf{x}, t)$ is needed. Only its existence is assumed.

Postulate 6. In continuum mathematics, even an infinitesimal part of the body can be considered as being a thermodynamic system.

4. THE FUNDAMENTAL INEQUALITY

In continuum mathematics, the Clausius inequality (2) can be written explicitly as

$$S(B) - S(A) > -\int_A^B dt \oint (\mathbf{q}/T) \cdot d\mathbf{a}, \tag{4}$$

where $\mathbf{q}$ is the heat flow through the surface of the body, $d\mathbf{a}$ its area element, and T the temperature of the heat bath applied at the area element $d\mathbf{a}$. This inequality holds according to the Postulates 5 and 6 also for an infinitesimal part of the body and can, if no diffusion occurs, be written by use of Gauss' law as

$$s(B) - s(A) > -\int_A^B dt(1/\rho) \nabla \cdot (\mathbf{q}/T). \tag{5}$$

Here, s is the specific entropy of the same element of matter, taken in the equilibrium states A and B.

We restrict ourselves now to fluids. By some simple transformations which involve the conservation laws for matter, momentum, and energy, and the differential relation

$$ds_{st} = (1/T_{st})\, du + (P_{st}/T_{st})\, d(\rho^{-1}), \tag{6}$$

one obtains the inequality

$$\int_{-\infty}^{\tau} dt\{(T_{st}^{-1} - T^{-1})\dot{u} + [(P_{st}/T_{st})\delta_{ik} - (P_{ik}/T)] \times \rho^{-1}(\partial v_i/\partial x_k) + \rho^{-1}\mathbf{q}\cdot\nabla T^{-1}\} \geq 0, \quad (7)$$

valid at all times τ for an arbitrary process in an element of matter which starts at $t = -\infty$ from an equilibrium state A. In its derivation, it is only assumed that the process from $t = -\infty$ to $t = \tau$ can be continued for $t > \tau$ in such a manner as to eventually reach an equilibrium state B with an integrand which vanishes for the continuation of the process from $t = \tau$ to $t = \infty$ in the limit of an infinitesimal part of the body.

The quantities T_{st}, P_{st}, s_{st} are the values of the temperature, the pressure, and the specific entropy as they were if with the values of the specific internal energy u and of the density ρ the state were an equilibrium state. They are therefore well-defined functions of u and ρ that can be taken from thermostatic tables and they are related by (6). The other quantities are the temperature T, the pressure tensor P_{ik}, and the velocity vector v_i.

We designate this inequality as the *fundamental inequality of thermodynamics* of processes because it makes a statement on the irreversible process which needs no unfounded nonequilibrium entropy, and because it is a substitute for the Clausius–Duhem inequality and leads to a more general thermodynamics of processes.

The fundamental inequality can also be formulated for deformable solid bodies and can be generalized so as to include diffusion and electromagnetic phenomena.

5. THE CONSTITUTIVE EQUATIONS

Constitutive equations reflect the properties of special materials. In complete thermostatic equilibrium, the properties of the fluid are expressible by functions of u and ρ. Such properties are the temperature T_{st}, the pressure P_{st}, and the entropy s_{st}. If, however, we consider a process which started from an equilibrium state at $t = -\infty$, then the values of the quantities T, P_{ik} and $\nabla(1/T)$ at time τ certainly depend uniquely upon how the element of matter has been treated during the interval $-\infty \leq t \leq \tau$. By treatment we mean here the supply of internal energy, the deformation, and the heat flow during the interval $-\infty < t \leq \tau$. Therefore, the quantities

$$T_{st}^{-1} - T^{-1}, \qquad \nabla(1/T), \qquad (P_{st}/T_{st})\mathbf{1} - (\mathbf{P}/T) \qquad (8)$$

at time τ ($\mathbf{1}$ = unit tensor, $\mathbf{P}$ = pressure tensor) are functionals

of u, of the heatflow, of the deformation (9)

over the interval $-\infty < t \leq \tau$, and besides they depend upon $\rho(\tau)$. A possible dependence upon $\rho(t)$ can be expressed by the deformation at previous times and by $\rho(\tau)$. Details are given in Ref. [5]. One postulate will, however, be explicitly stated. It is the property of approach to equilibrium. By this it is meant that these functionals tend to zero and consequently the quantities in (8) tend also to zero for $t \to \infty$ if from some time on $\dot{u}$, $\mathbf{q}$, and $\boldsymbol{\eta}$ are equal to zero. It is obvious that there are two different kinds of constitutive equations. The first set of constitutive equations is concerned with the equilibrium properties of a material. They can be condensed into one function—viz., a thermodynamic potential. As such, one can take the static entropy s_{st} as a function of the variables u and ρ in the case of fluids, which is considered here for the sake of demonstration. From the differential relation one has therefore the *static constitutive equations*

$$s_{st} = s_{st}(u, \rho), \qquad T_{st} = T_{st}(u, \rho), \qquad P_{st} = P_{st}(u, \rho),$$
$$c_v = c_v(u, \rho), \qquad \kappa = \kappa(u, \rho), \cdots, \text{etc.} \tag{10}$$

where c_v = specific heat at constant volume, κ = isothermal or adiabatic compressibility. The principle of equipresence [6] is here quite naturally satisfied.

The second set of constitutive equations expresses the quantities in (8) in terms of the quantities indicated in (9). These relations properly should be called dynamic constitutive equations. If all quantities in (8) are permitted to depend to the full extent upon the quantities (9) over the whole interval $-\infty < t \leq \tau$, unless this dependence contradicts some physical law or principle, then the principle of equipresence is satisfied. For special materials, or for a special class of processes, it may be that it suffices to permit a restricted history, but then it should be restricted in the same way in all dynamic constitutive equations. An example of restricted histories is given in the next section.

In the theory presented here, there is no principle of equipresence which applies simultaneously to the static and to the dynamic constitutive equations.

6. CLASSICAL THERMODYNAMICS OF IRREVERSIBLE PROCESSES

The constitutive equations as described here are of the after-effect type. They have to be such that for all processes the fundamental inequality (7) holds. Moreover, they have to satisfy the principle of material frame indifference [7, 8] (for its history see Ref. [9]) and express the isotropy of the fluid.

We do not give the restrictions imposed by these conditions, but we shall discuss an important special case of these constitutive equations.

Assume that the mentioned functionals depend upon $u(\tau)$ and $\rho(\tau)$, and that the previous treatment enters them only by the time derivatives of the specific internal energy and of the deformation gradient and, furthermore, by the heat flow $\mathbf{q}$, all of them taken at time τ, then the functionals become ordinary functions and the principle of material frame indifference and the isotropy of the fluid require the special forms of these functions

$$T_{st}^{-1} - T^{-1} = F_0(u, \rho; \dot{u}, \mathbf{q}, \boldsymbol{\eta}), \tag{11}$$

$$\nabla(1/T) = \mathbf{F}_1(u, \rho; \dot{u}, \mathbf{q}, \boldsymbol{\eta}), \tag{12}$$

$$(P_{st}/T_{st}) - (\mathbf{P}/T) = \mathbf{F}_2(u, \rho; \dot{u}, \mathbf{q}, \boldsymbol{\eta}), \tag{13}$$

where F_0 is a scalar-valued function, $\mathbf{F}_1$ a vector-valued function, and $\mathbf{F}_2$ a symmetric tensor valued function* of the vector $\mathbf{q}$ and of the symmetric tensor $\boldsymbol{\eta}$. This tensor is defined as

$$\eta_{ik} = \tfrac{1}{2}(\partial v_i/\partial x_k + \partial v_k/\partial x_i). \tag{14}$$

A fluid of this type is termed as being of the differential type and of complexity 1 [10]. The property of approach to equilibrium implies that the right members of (11), (12), and (13) vanish if $u = 0$, $\mathbf{q} = 0$, and $\boldsymbol{\eta} = 0$.

The fundamental inequality imposes a further restriction on these constitutive equations, which has been made explicit by J. Keller (Aachen) (see Appendix in Ref. [5]). It states that not only the integral in the fundamental inequality, but the integrand itself is non-negative at all times if the constitutive equations are of the form (11), (12), (13).

The identity

$$\rho(ds_{st}/dt) + \nabla(\mathbf{q}/T) = (T_{st}^{-1} - T^{-1})\rho(du/dt) + \mathbf{q}\nabla(1/T) + \operatorname{Tr}[(P_{st}/T_{st})\mathbf{1} - (\mathbf{P}/T)]\boldsymbol{\eta} \tag{15}$$

which is a combination of (6) and of the energy law and of the continuity equation, and which has been used in the transition from (5) to (7), now gains increased importance. The right member is the integrand of the fundamental inequality, apart from a factor ρ, and is never negative for materials of the differential type and of complexity 1. Consequently, this equation can be interpreted as an entropy balance equation with $s_{st}(u, \rho)$ as the specific entropy during the process, $\mathbf{q}/T$ as the entropy flow, and with the right member as the entropy production Therefore, for such materials, the thermostatic entropy also plays the part of a possible entropy function during the process. Moreover, the constitutive equations (10) to (12) are just the phenomenological equations or rate equations of classical thermo-

* Herewith we assume that there is no exchange of intrinsic angular momentum and moment of momentum of the motion.

dynamics of irreversible processes. There is only one difference. In classical thermodynamics of irreversible processes, there never has been a distinction between T_{st} and T and, therefore, one of the constitutive equations dropped out. But this distinction could have been made already in the classical theory with just the same argument with which it distinguished between the dynamic pressure tensor $\mathbf{P}$ and the static pressure $P\mathbf{1}$.

7. COLEMAN'S THERMODYNAMICS OF SIMPLE MATERIALS

We shall not discuss here the position of Coleman's general thermodynamic theory [4] within the framework of our thermodynamic theory. Rather we restrict ourselves to confronting the special case investigated by Coleman and Mizel [3] with our results on simple materials of the differential type and of complexity 1 because there is a strong formal similarity between the mutual results.

Coleman and Mizel start with the set of constitutive equations (so called constitutive assumption)

$$s = s(u, \mathbf{q}, \mathbf{F}, \dot{\mathbf{F}}), \tag{16}$$

$$T = T(u, \mathbf{q}, \mathbf{F}, \dot{\mathbf{F}}), \qquad \mathbf{P} = \mathbf{P}(u, \mathbf{q}, \mathbf{F}, \dot{\mathbf{F}}), \qquad \text{grad}\, T^{-1} = X(u, \mathbf{q}, \mathbf{F}, \dot{\mathbf{F}}), \tag{17}$$

in which we have somewhat changed the independent and the dependent variables in order to facilitate the comparison with our formulas (11), (12), (13) and the first formula in (10). There is also a difference in that Coleman and Mizel's constitutive equations are written for deformable solid materials while ours apply to fluids; therefore, the former contain the deformation gradient $\mathbf{F}$ and its material time derivative $\dot{\mathbf{F}}$ in place of ρ and $\boldsymbol{\eta}$.

The main result of Coleman and Mizel is that, by virtue of the Clausius–Duhem inequality, the entropy s and the temperature T cannot depend upon $\mathbf{q}$ and $\dot{\mathbf{F}}$ although such a dependence would be in line with the principle of equipresence. So, within the class of constitutive equations (16) and (17), they have proved the existence of a caloric equation of state $s = s(u, \mathbf{F})$ and of a thermal equation of state (temperature relation) $T = T(u, \mathbf{F})$ during a process.

We obtain the same result directly from (15), since we have already proved the existence of an entropy $s_{st}(u, \rho)$ which satisfies an inequality of the Clausius–Duhem type

$$\rho(ds_{st}/dt) + \nabla(\mathbf{q}/T) \geq 0,$$

and, applying similar arguments as used by Coleman and Mizel, we arrive at $T = T(u, \mathbf{F})$.

We have only to assume that there is no distinction between T and T_{st}, which makes $F_0(u, \mathbf{F}; \dot{u}, \mathbf{q}, \dot{\mathbf{F}}) = 0$ and, therefore, makes $\dot{u}$ a dependent variable, so that now our constitutive equations (12) and (13) have the same general form as the second and third constitutive equations in (17). We could instead assume directly that the right members in (11), (12), (13) do not depend upon $\dot{u}$. Then, from the non-negative value of the entropy production [i.e. the right member in (15)], one concludes again that $T = T_{st}(u, \mathbf{F})$.

We believe, however, that the Coleman and Mizel constitutive equations (16) and (17) are assumed in too special a form. Although they use the principle of equipresence in the variables u, $\mathbf{q}$, $\mathbf{F}$, $\dot{\mathbf{F}}$, it is a question whether this set of variables is a consistent or a uniform or, in some respect, a complete set. In equilibrium, we have only u and $\mathbf{F}$ as independent variables. In nonequilibrium, one can excogitate many new independent variables and it is certainly justified to ask what selection should be made on a certain level of description. We would like to supplement the principle of equipresence by the requirement of uniform complexity and might then speak of the *strong principle of equipresence*. By uniform complexity we mean, in our case, that the constitutive equations for the quantities in (8) depend, besides u and $\mathbf{F}$, upon the derivatives of $\dot{u}$, $\mathbf{q}$, and $\dot{\mathbf{F}}$ (in the case of deformable solids) up to the same order $r - 1$. Such a material we designate to be of the differential type and of order r. For $r = 1$, we recover our constitutive equations (11), (12), (13).

If we apply this strong principle of equipresence to Coleman and Mizel's constitutive equations (16) and (17), then the variables s, T, $\mathbf{P}$, grad $1/T$ also have to depend upon $\dot{u}$ if a dependence upon $\mathbf{q}$ and $\mathbf{F}$ is admitted. In this case, Coleman and Mizel's procedure again leads to a caloric equation of state $s = s(u, \mathbf{F})$ even during a process, but since T may depend upon $\dot{u}$, it can no longer be inferred that $T^{-1} = \partial s(u, \mathbf{F})/\partial u$. Rather, the constitutive equations for T, grad $1/T$, and for the stress tensor agree with our constitutive equations (11), (12), (13) and are subject to the same restrictions by the non-negative value of the entropy production. It is a minor point in this discussion that Coleman and Mizel use T as an independent variable; this would only imply that $\dot{T}$ should be included in their set of independent variables.

We consider it as an interesting result, that the constitutive equations (17) with $\dot{u}$ as an additional independent variable, or equivalently the constitutive equations (11), (12), (13) lead to the existence of an entropy function $s_{st} = s_{st}(u, \mathbf{F})$ with the growth property, without the assumption that an entropy exists and without the use of a Clausius–Duhem inequality. Both, the existence of an entropy which depends upon u and $\mathbf{F}$ only, even during a process and the Clausius–Duhem inequality, are a mathematical result derived from the fundamental inequality for materials of the differential type and of complexity 1.

8. GENERALIZATIONS

The theory which has been sketched here may be modified and generalized in various directions. When we speak of modification, we mean the corresponding theory for deformable solid materials. This theory has already been worked out [11].

A further task is the formulation of the theory for materials in which diffusion of components must be taken into account. This can be done without essential difficulty.

Another problem is the formulation of the theory on a higher level of description, that means a description in which not only the histories of u, of q, and of the deformation gradient $\mathbf{F}$ enter, but in which internal variables are taken into account explicitly. This should bring about constitutive equations with a shorter memory than in the lower-level description.

One might also be interested in a certain classification of the constitutive functionals—as of the differential type of complexity higher than 1—or of the rate type or of the integral type or of the mixed integral-rate type [10].

It would be particularly interesting whether, or to what extent, such materials can be explained as materials of simpler types—for instance, of the differential type and of complexity r or of the rate type on a higher level of description. Only in the linear domain—i.e., for small departures from a fixed reference state—are some results available [1, 12].

REFERENCES

1. J. Meixner, Beziehungen zwischen Netzwerktheorie und Thermodynamik, Sitzungsberichte der Arbeitsgemeinschaft für Forschung des Landes Nordrhein-Westfalen, Köln und Opladen, Westdeutscher Verlag, 1968.
2. F. Schlögl, Informationstheorie und Thermodynamik irreversibler Prozesse, *Ibid.*, 1968.
3. B. D. Coleman and V. Mizel, J. Chem. Phys. *40*, 1116 (1964).
4. B. D. Coleman, Arch. Rational Mech. Analysis *17*, 1, 230 (1964).
5. J. Meixner, Z. Physik, *219*, 79 (1969).
6. C. Truesdell and W. Noll: The non-linear field theories of mechanics, *in* Encyclopedia of Physics, Bd. III/3, p. 359 (S. Flügge, Ed.), Springer-Verlag, 1965.
7. J. G. Oldroyd, Proc. Roy. Soc. (London) *A200*, 523 (1950).
8. R. S. Rivlin and J. L. Ericksen, J. Rational Mech. Analysis *4*, 323 (1955).
9. P. 45–47 of reference 6.
10. See Ref. 6, p. 92.
11. J. Meixner, Arch. Rational Mech. Analysis *33*, 33 (1969).
12. J. Meixner, Z. Physik *193*, 366 (1966).

Comments

R. E. Barieau: How do you define the temperature and the pressure for a system that is in nonequilibrium?

J. Meixner: May I ask, do you mean the static temperature and the static pressure?

R. E. Barieau: Your T is temperature. I understand what $1/\Pi$ means, but how do you find the T? You referred to it as the real temperature.

J. Meixner: That is a very good question, I would say. It has been asked many times, and I have never found a really very good answer for it.

But, first, I would say, I have done away with entropy. The next step might be to let us also do away with temperature. But, I am afraid this does not give you very much. To be more serious:

There are two definitions possible for temperature. The first would be, we understand as temperature what is defined by the translational degrees of freedom. This means one is close to a Maxwell distribution of the velocities of the molecules, and define T by this distribution. Another definition, a little too lengthy to give here in full, presumes the existence of heat baths with infinitely small relaxation times.

B. Robertson: Temperature, entropy, and other thermodynamic variables for systems not in equilibrium have been defined in a paper of mine in *The Physical Review,* wherein I give reference to earlier work by Jaynes and Richardson.

This definition of entropy is unique, once the system is specified and once the experiment that is being done on the system is specified.

J. Meixner: I would be very interested to read your paper, and I would be grateful if you could give me the reference.

However, there is a simple answer. You may be able to define entropy in nonequilibrium; however, this is not an objective quantity, but also depends on the observer, and his knowledge about the system, and how fully he wants to describe it.

B. Robertson: The definition I gave (Phys. Rev. *144*, 151; *160*, 175) is not subjective. Once the observer's experiment is specified, the entropy is uniquely determined.

J. Meixner: Then you need something like a complete set of experiments. I would not object to that, but you have always an incomplete set of experiments, for instance, when you apply just heat baths, and a distribution of forces on the surface of the system, and you observe the heat flow going into the system, and let's say how the system is going into motion. Well, if you analyze the relation between these quantities that is also thermodynamics. But this is a description which does not give you very much information about what is going on inside, and certainly it does not permit one to assign a unique entropy to the system in nonequilibrium states.

AUTHOR CITATION INDEX

SUBJECT INDEX

About the Editor

JOSEPH KESTIN is Professor of Engineering at Brown University. He teaches courses in classical and statistical thermodynamics, heat transfer, and boundary layer theory, and is engaged in research on turbulence effects in heat transfer and the transport properties of gases and ionic solutions. He was the Head of the Department of Mechanical Engineering at the Polish University College in London.

Professor Kestin graduated with a degree in mechanical engineering from the Polytechnic Institute in Warsaw, Poland, earned the Ph.D. at Imperial College of the University of London, and in 1966 obtained the degree of Doctor of Science from London University. He has held visiting professorships at Imperial College, at the University of Paris (Sorbonne), and at Université Claude Bernard in Lyon, France. He was a Fulbright Lecturer at the Instituto Superior Técnico in Lisbon, Portugal, and served as advisor to the Chancellor of the University of Teheran, Iran.

Professor Kestin served on several panels and committees of the National Research Council of the National Academy of Sciences. Currently he is Chairman of the Evaluation Panel for the Office of Standard Reference Data of the National Bureau of Standards, a member of the Executive Committee of the Evaluation Panels for NBS, and a member of the Numerical Data Advisory Board of NAS. He held numerous offices at the American Society of Mechanical Engineers and was Chairman of the Applied Mechanics Division as well as Technical Editor of the *Journal of Applied Mechanics*. Currently he is a member of the National Nominating Conference of ASME and President of the International Association for the Properties of Steam.

Professor Kestin has published two books on thermodynamics, translated three books on thermodynamics and boundary layer theory, and written several contributing sections to handbooks and encyclopedias.